Lecture Notes in Mathematics

Edited by A. Dold and B. Eckmann

531

Yau-Chuen Wong

The Topology of Uniform Convergence on Order-Bounded Sets

Springer-Verlag
Berlin · Heidelberg · New York 1976

Author
Yau-Chuen Wong
Department of Mathematics
United College
The Chinese University of Hong Kong
Shatin, N.T./Hong Kong

Library of Congress Cataloging in Publication Data

Wong, Yau-chuen.
The topology of uniform convergence on order-bounded sets.

(Lecture notes in mathematics ; 531)
Bibliography: p.
Includes index.
1. Linear topological spaces. 2. Convergence.
3. Duality theory (Mathematics) I. Title. II. Series: Lecture notes in mathematics (Berlin) ; 531.
QA3.L28 vol. 531 [QA322] 515'.73 76-26481

AMS Subject Classifications (1970): 06A65, 46A05, 46A15, 46A20, 46A35, 46A40, 46A45, 46A99, 47B55, 47D15

ISBN 3-540-07800-2 Springer-Verlag Berlin · Heidelberg · New York
ISBN 0-387-07800-2 Springer-Verlag New York · Heidelberg · Berlin

Printing and binding: Beltz Offsetdruck, Hemsbach/Bergstr.

CONTENTS

INTRODUCTION

In studying ordered topological vector spaces, particularly important roles are played by two intrinsic topologies: the order-bounded (or order) topology and the topology σ_S of uniform convergence on all order-bounded sets. The order-bound topology was studied independently by Schaefer [1] and Namioka [1] , while the topology σ_S was studied by Nakano [1] and Dieudonné in the special case of locally convex Riesz spaces, and by Peressini [3] in a fairly general setting (he used the notation $o(E, E')$). A remarkable theorem of Nakano [1] (asserting that, for topological Riesz spaces, topological completeness follows from certain order completeness assumption) is one of the deepest results in the theory of locally convex Riesz spaces; the author showed, in 1969, that σ_S is relevant for establishing a converse of Nakano's theorem. Therefore it is interesting to seek some necessary and sufficient condition for a given locally solid topology $\mathscr{P}$ on E to be $o(E, E')$. One of the purposes of these lecture notes is to give such characterizations by means of some special classes of continuous linear mappings, and another purpose is an attempt to provide a unifying treatment of nuclear spaces and the topology $o(E, E')$. The guiding concepts in this approach are those of absolutely summing mappings and cone-absolutely summing mappings. These concepts are studied only in the general setting here; for the speciality of such an account in the Banach lattices setting the reader is referred to the excellent book written by Schaefer [3] .

The first chapter is a brief discussion of duality problems for ordered vector spaces, and of the constructions of the topologies $\mathscr{P}_F$, $\mathscr{P}_D$ and $\mathscr{P}_S$ (respectively the locally o-convex topology, the locally decomposable topology and the locally solid topology associated with $\mathscr{P}$). The second chapter mainly deals with some useful classes of locally solid topologies on certain vector subspaces of E^A, and studies the corresponding dual structures. The final chapter is devoted to a study of cone-absolutely summing mappings, of cone-prenuclear mappings and of the topology $o(E, E')$.

Throughout these notes, (i, j, k) will denote the k-th proposition or theorem in Chapter i, Section j.

Parts of these notes were delivered at Yale University, during the period from September 1973 to March 1974, and at McMaster University in Canada, during the period of April to June 1974. The material in these notes is based on Seminar Report 'Lecture notes on nuclear and L-nuclear spaces' published by Yale University, 1974. Unfortunately there are some errors and misprints there, I would like to apologize for this, and take the opportunity to put them right here.

The author would like to thank Yale-in-China Association for financial and moral support during his stay in New Haven. His stay at Yale University was very agreeabl and the friendly atmosphere and the working conditions were conducive to scientific works.

I am much indebted to Professors S. Kakutani (Yale) and T. Husain (McMaster) for interesting discussions and valuable suggestions.

Finally, I would like to thank Mr. Billy P.M. Lam who expertly carried out the task of typing the manuscript.

Y.C.W.

United College
The Chinese University of Hong Kong
November 1975

CHAPTER 1. A SURVEY OF ORDERED VECTOR SPACES

In this chapter we review some basic concepts and propositions in ordered vector spaces which we shall need in what follows. The Hahn-Banach extension theorem of ordered type, given in Theorem (1.1.1), is due to Bonsall [1], and is extremely useful for establishing the duality problems of ordered vector spaces. The first duality theorem (1.1.7), concerned with the order-convexity and decomposability, has many important applications; for instance, Schaefer's duality theorem for normal and b-cones, which is a generalization of Krein-Grosberg's duality theorem for α-normal and α-generating, is deduced. A slight generalization of Andô-Ellis' duality theorem for α-generating and $(\alpha+\varepsilon)$-normal is given in the final section. The second duality theorem (1.1.9), due to Jameson [1], is concerned with the absolute order-convexity and absolutely dominated property. In the final section, we study the construction of $\mathcal{P}_F$, due to Namioka [1], and of $\mathcal{P}_D$, due to Wong and Cheung [1], also study the relationship between $\sigma_S(E, E')$ and the topology $o(E, E')$, due to Peressini [1], of uniform convergence on order-intervals in detail.

Terminology and notation concerning ordered vector spaces will follow Schaefer [3], Peressini [1] and Wong and Ng [1], while Köthe [1] and Schaefer [1] will serve as our references for material on topological vector spaces. The background material concerning absolutely summing mappings can be found in Pietsch [1].

1.1. Duality theorems

Throughout these notes, the scalar field for vector spaces is assumed to be the field $\mathbb{R}$ of real numbers. By a (positive) cone C in a vector space E is meant a non-empty convex subset of E which satisfies $\lambda C \subseteq C$ for all $\lambda \geq 0$. A cone C is proper if $C \cap (-C) = \{0\}$. Clearly a cone C in E determines a transitive and reflexive relation "$\leq$" by

$$x \leq y \quad \text{if} \quad y - x \in C ;$$

moreover this relation is compatible with the vector structure, i.e.,

(a) if $x \geq 0$ and $y \geq 0$ then $x + y \geq 0$,

(b) if $x \geq 0$ then $\lambda x \geq 0$ for all $\lambda \geq 0$.

The relation determined by the cone is called the vector ordering, and the pair (E, C) (or $(E, \leq)$) is referred to as an ordered vector space. It is also clear that a cone C is proper if and only if the vector ordering $\leq$, induced by C, is antisymmetric, i.e., if $x \geq 0$ and $x \leq 0$ then $x = 0$.

Let (E, C) be an ordered vector space with the positive cone C. Denote by E^* the algebraic dual of E, and by C^* the dual cone of C, that is $C^* = \{f \in E^* : f(u) \geq 0 \text{ for all } u \in C\}$; elements in C^* are called positive linear functionals on E. Let p be a functional defined on C. p is said to be sublinear if it satisfies the following conditions:

$$p(x + y) \leq p(x) + p(y) \quad \text{and} \quad p(\lambda x) = \lambda p(x)$$

whenever $x, y \in C$ and $\lambda \geq 0$; and p is said to be superlinear if $-p$ is sublinear. In studying the duality problems for ordered vector spaces, the following theorem, due to Bonsall [1], is very useful.

(1.1.1) Theorem (Bonsall). Let (E, C) be an ordered vector space and p a sublinear functional defined on E. Suppose further that q is a superlinear functional defined on C such that

$$q(u) \leq p(u) \quad \text{for all} \quad u \in C .$$

Then there exists a linear functional g on E such that

$$q(u) \leq g(u) \quad \text{for all} \quad u \in C$$

and

$$g(x) \leq p(x) \quad \text{for all} \quad x \in E .$$

Proof. Define, for each $x \in E$, that

$$r(x) = \inf \{p(x + u) - q(u) : u \in C\} .$$

It is not hard to see that r is a (finite) sublinear functional on E for which

$$r(x) \leq p(x) \qquad (x \in E)$$

and

$$r(-u) \leq -q(u) \quad (u \in C).$$

Applying the Hahn-Banach extension theorem to get a linear functional g on E which is dominated by r. It is then easily seen that this linear functional g on E has the required properties.

As an application of Bonsall's theorem, we prove a sum-theorem, which is analogues to the Riesz decomposition property.

(1.1.2) Corollary. Let $p_1, \ldots, p_n$ be sublinear functionals defined on a vector space X, let Y be a vector subspace of X and suppose that f is a linear functional defined on Y such that

$$f(y) \leq \sum_{i=1}^{n} p_i(y) \quad \text{for all} \quad y \in Y .$$

<u>Then there exist linear functionals</u> $f_1, f_2, \ldots, f_n$ <u>on</u> X <u>such that</u>

$$f(y) = \sum_{i=1}^{n} f_i(y) \quad \underline{\text{for all}} \quad y \in Y$$

<u>and</u> $$f_i(x) \leq p_i(x) \quad \underline{\text{for all}} \quad x \in X \quad \underline{\text{and}} \quad i = 1, 2, \ldots, n .$$

<u>Proof</u>. We verify the case of $n = 2$, and then the proof can be completed by induction. We first note that Y is a cone in X . Since

$$f(y) - p_2(y) \leq p_1(y) \quad \text{for all} \quad y \in Y$$

and since the restriction of $f - p_2$ on Y is superlinear, it follows from Bonsall's theorem that there exists a linear functional f_1 on X such that

$$f(y) - p_2(y) \leq f_1(y) \quad \text{for all} \quad y \in Y$$

and $$f_1(x) \leq p_1(x) \quad \text{for all} \quad x \in X .$$

Since the restriction of $f - f_1$ on Y is linear and satisfies

$$f(y) - f_1(y) \leq p_2(y) \quad \text{for all} \quad y \in Y ,$$

applying Bonsall's theorem again to get a linear functional f_2 on X such that

$$f(y) - f_1(y) \leq f_2(y) \quad \text{for all} \quad y \in Y \tag{1.1}$$

and $$f_2(x) \leq p_2(x) \quad \text{for all} \quad x \in X .$$

Since Y is a vector subspace of X and since $f - f_1$ is a linear functional on Y , it follows from the Formula (1.1) that

$$f(y) - f_1(y) = f_2(y) \quad (y \in Y) .$$

As another application of Bonsall's theorem, we prove a positive extension theorem, due to Namioka [1] and Bauer [1].

(1.1.3) Theorem (Namioka and Bauer). Let (E, C) be an ordered vector space, G a subspace of E and suppose that f is a linear functional defined on G. Then the following statements are equivalent:

(a) f can be extended to a positive linear functional on E.

(b) There exists a convex absorbing subset V of E such that

$$f(y) \leq 1 \quad \text{for all} \quad y \in G \cap (V - C) .$$

Proof. The implication (a)$\Rightarrow$(b) is clear; in fact, if g is a positive extension of f on E, then the set

$$V = \{x \in E : g(x) \leq 1\}$$

has the desired property. To prove the implication (b)$\Rightarrow$(a), we note that $V - C$ is convex and absorbing, and that the gauge of $V - C$, denoted by p, is a sublinear functional on E such that

$$-C \subset p^{-1}(0) \quad \text{and} \quad f(y) \leq p(y) \quad \text{for all} \quad y \in G .$$

G is a cone in E, and f is a superlinear functional on G; by Bonsall's theorem, there exists a linear functional g on E such that

$$f(y) \leq g(y) \quad \text{for all} \quad y \in G \tag{1.2}$$

and

$$g(x) \leq p(x) \quad \text{for all} \quad x \in E . \tag{1.3}$$

Since G is a subspace of E and since f is linear on G, it follows from Formula (1.2) that $f(y) = g(y)$ for all $y \in G$. On the other hand, since $-C \subset p^{-1}(0)$, we conclude from Formula (1.3) that $g(u) \geq 0$ for all $u \in C$; therefore g is a positive extension of f on E.

Let $\langle E, F \rangle$ be a dual pair. If V is a subset of E, the polar of V, taken in F, is defined by

$$V^{\circ} = \{f \in F : \langle x, f \rangle \leq 1 \quad \text{for all} \quad x \in V\} .$$

Since $\langle E, E^* \rangle$ is always a dual pair, the polar of V, taken in E^*, is denoted by V^{π}. Throughout this book, $\tau(E, F)$ (resp. $\sigma(E, F)$ and $\beta(E, F)$) denotes the Mackey (resp. weak and strong) topology on E.

(1.1.4) Lemma. *Let* $\langle E, F \rangle$ *be a duality and suppose that* $\{B_\alpha : \alpha \in D\}$ *is a family of convex subsets of* E, *each containing* 0. *Then* $\bigcap_\alpha \overline{B}_\alpha = \overline{\bigcap_\alpha B_\alpha}$ *if and only if*

$$(\bigcap_\alpha B_\alpha)^\circ = \overline{co}\,(\bigcup_\alpha B_\alpha^\circ) ,$$

where the upper-bars denote the weak-*closures. In particular, if each* B_α *is a* $\tau(E, F)$-*neighbourhood of* 0, *then* $\bigcap_\alpha \overline{B}_\alpha = \overline{\bigcap_\alpha B_\alpha}$.

Proof. Necessity. By the bipolar theorem and the hypotheses, we have

$$(\bigcap_\alpha B_\alpha)^\circ = (\overline{\bigcap_\alpha B_\alpha})^\circ = (\bigcap_\alpha \overline{B}_\alpha)^\circ = (\bigcap_\alpha B_\alpha^{\circ\circ})^\circ$$

$$= (\bigcup_\alpha B_\alpha^\circ)^{\circ\circ} = \overline{co}\,(\bigcup_\alpha B_\alpha^\circ) .$$

Sufficiency. We first note that $\bigcap_\alpha B_\alpha$ is convex and contains 0, it then follows from the bipolar theorem that

$$\bigcap_\alpha \overline{B}_\alpha = \bigcap_\alpha B_\alpha^{\circ\circ} = (\bigcup_\alpha B_\alpha^\circ)^\circ = (\overline{co}\,(\bigcup_\alpha B_\alpha^\circ))^\circ$$

$$= (\bigcap_\alpha B_\alpha)^{\circ\circ} = \overline{\bigcap_\alpha B_\alpha} .$$

Finally, suppose that each B_α is a $\tau(E, F)$-neighbourhood of 0 and that $x \in \bigcap_\alpha \overline{B}_\alpha$. As 0 is an interior point of each B_α, it follows from Schaefer [1, (II.1.1)] that λx is an interior point of B_α for all $\lambda \in [0, 1)$ and $\alpha \in D$, and hence that $\lambda x \in \bigcap_\alpha B_\alpha$ for all $\lambda \in [0, 1)$. Letting $\lambda \to 1$, we obtain $x \in \overline{\bigcap_\alpha B_\alpha}$.

In particular, if V and W are convex $\tau(E, F)$-neighbourhoods of 0, it follows from the Alaoglu-Bourbaki theorem and Schaefer [1, (II.10.2)] that $(V \cap W)^\circ = \text{co}(V^\circ \cap W^\circ)$.

Let (E, C) and (F, K) be ordered vector spaces and let $\langle E, F\rangle$ be a dual pair. If $K = -C^\circ$, then we say temporarily that $\langle E, F\rangle$ is an <u>ordered duality on the right</u>. Ordered duality on the left is defined dually; while an ordered duality is defined to be ordered duality on both right and left.

In order to establish duality theorems for ordered vector spaces, we present the following lemma which is the Key to the duality problem.

(1.1.5) Lemma. <u>Let</u> (E, C) <u>and</u> (F, K) <u>be ordered vector spaces which forms an ordered duality on the right</u>, <u>and let</u> V <u>be a subset of</u> E. <u>Then the following assertions hold</u>:

(a) <u>If</u> $0 \in V$ <u>then</u>

$$(V + C)^\circ = V^\circ \cap C^\circ \quad \text{and} \quad (V - C)^\circ = V^\circ \cap (-C^\circ) ; \tag{1.4}$$

<u>if</u>, <u>in addition</u>, V <u>is symmetric then</u>

$$(V + C)^\circ = -(V^\circ \cap K) \quad \text{and} \quad (V - C)^\circ = V^\circ \cap K . \tag{1.5}$$

(b) <u>If</u> V <u>is a convex</u> $\tau(E, F)$-<u>neighbourhood of</u> 0, <u>then</u>

$$(V \cap C)^\circ = V^\circ + C^\circ = V^\circ - K ; \tag{1.6}$$

<u>if</u>, <u>in addition</u>, V <u>is symmetric</u>, <u>then</u>

$$(-(V \cap C))^\circ = V^\circ + K . \tag{1.7}$$

<u>Proof</u>. (a) As V and C are subsets of $V + C$, we have $(V + C)^\circ \subset V^\circ \cap C^\circ$; conversely if $f \in V^\circ \cap C^\circ$, then for any $x \in V$ and

$v \in C$, we have

$$\langle x + v, f\rangle = \langle x, f\rangle + \langle v, f\rangle \leq \langle x, f\rangle \leq 1 ,$$

which implies that $f \in (V + C)^{\circ}$. Therefore $(V + C)^{\circ} = V^{\circ} \cap C^{\circ}$. The other equality in (1.5) can be verified in the same way.

(b) Since each element in C° is negative on C , it follows that $V^{\circ} + C^{\circ} \subseteq (V \cap C)^{\circ}$. To prove the opposite inclusion, let $f \in (V \cap C)^{\circ}$ and suppose that p is the gauge of V . Then p is a $\tau(E, F)$-continuous sublinear functional on E such that

$$f(u) \leq p(u) \quad \text{for all} \quad u \in C .$$

The restriction of f to C is a superlinear; by Bonsall's theorem, there exists a linear functional g on E such that

$$f(u) \leq g(u) \quad \text{for all} \quad u \in C$$

$$g(u) \leq p(u) \quad \text{for all} \quad x \in E .$$

Since $\tau(E, F)$ is consistent with $\langle E, F\rangle$ and since p is $\tau(E, F)$-continuous, it follows that $g \in F$, and hence that $g \in V^{\circ}$. Clearly $g - f \in K = -C^{\circ}$, therefore we have $f = g - (g - f) \in V^{\circ} - K = V^{\circ} + C^{\circ}$ as required.

Finally, if V is symmetric, we then have

$$(-(V \cap C))^{\circ} = -(V \cap C)^{\circ} = -(V^{\circ} - K) = V^{\circ} + K .$$

Before giving duality theorems, we introduce the following notation. For any subset V of an ordered vector space (E, C) , we define

$$F(V) = (V + C) \cap (V - C) ;$$

$$D(V) = \{x \in V : x = \lambda x_1 - (1 - \lambda)x_2 \ , \lambda \in [0, 1], x_1, x_2 \in V \cap C\} ;$$

$$B(V) = \cup\{[-u, w] : u, w \in V \cap C\} ;$$

$$S(V) = \cup\{[-u, u] : u \in V \cap C\} .$$

Then $F(V)$ is called the order-convex hull of V, while $D(V)$ is called the decomposable kernel of V.

(1.1.6) Lemma. *For subsets* V *and* W *of* (E, C), *the following assertions hold:*

(a) *if* V *is convex and symmetric, then*

$$D(V) = co\{-(V \cap C) \cup (V \cap C)\} ;$$

(b) $D(V) \cap C = V \cap C \subset S(V) \cap C = B(V) \cap C$; *if, in addition,* $0 \in V$, *then*

$$B(V) \cap C \subset F(V) \cap C ;$$

(c) *if* $V \subset W$ *then* $D(V) \subset D(W)$, $F(V) \subset F(W)$, $S(V) \subset S(W)$ *and* $B(V) \subset B(W)$;

(d) *if* V *is symmetric then* $-(V \cap C) \subset D(V)$ *and*

$$B(V) = F(D(V)) = \{-(V \cap C) + C\} \cap \{(V \cap C) - C\} \subset F(V) ;$$

(e) *if* V *is convex then*

$$D(F(V)) \subset S(V) ;$$

(f) *if* V *is convex and symmetric then*

$$D(V) \subset D(F(V)) \subset S(V) \subset B(V) \subset F(V) .$$

Proof. (a) Obvious.

(b) It is clear that

$$V \cap C \subset D(V) \cap C \subset V \cap C \subset S(V) \cap C \subset B(V) \cap C \subset S(V) \cap C ;$$

therefore we get the desired result.

(c) Obvious.

(d) Let $u \in V \cap C$. Since V is symmetric, $-u \in V$, we conclude from

$$-u = 0 \cdot y - (1 - 0)u$$

with $y \in V \cap C$ that $-u \in D(V)$. This shows that $-(V \cap C) \subset D(V)$.

If $x \in B(V)$, then there exist $u, w \in V \cap C$ such that $-u \leq x \leq w$. It then follows from $-u, w \in D(V)$ that $x \in F(D(V))$, and thus $B(V) \subset F(D(V))$.

On the other hand, if $x \in F(D(V))$ then there exist y_1, y_2, z_1, z_2 in $V \cap C$ and $\lambda, \mu \in [0, 1]$ such that

$$\mu z_1 - (1 - \mu) z_2 \leq x \leq \lambda y_1 - (1 - \lambda) y_2 ,$$

clearly

$$-z_2 \leq x \leq y_1 .$$

Therefore $x \in B(V)$ and $x \in \{-(V \cap C) + C\} \cap \{(V \cap C) - C\}$. This shows that $F(D(V)) = B(V) \subset \{-(V \cap C) + C\} \cap \{(V \cap C) - C\}$. Since $-(V \cap C)$ and $(V \cap C)$ are in $D(V)$, it follows that

$$\{-(V \cap C) + C\} \cap \{V \cap C) - C\} \subset F(D(V)) .$$

Finally, since $D(V) \subset V$, we conclude that $F(D(V)) \subset F(V)$.

(e) If $x \in D(F(V))$, then there exist $\lambda \in [0, 1]$, $u_1, u_2 \in F(V) \cap C$ such that

$$x = \lambda u_1 - (1 - \lambda) u_2 .$$

Let w_1, w_2, in $V \cap C$, be such that $0 \leq u_i \leq w_i$ $(i = 1, 2)$. The convexity of V ensures that $w = \lambda w_1 + (1 - \lambda) w_2 \in V \cap C$. Clearly $\pm x \leq w$, we conclude that $x \in S(V)$ which implies that $D(F(V)) \subset S(V)$.

(f) This follows from (c), (d) and (e) .

Let V be a subset of (E, C) . V is said to be

(a) <u>order-convex</u> if $V = F(V)$;

(b) <u>o-convex</u> if it is both order-convex and convex;

(c) <u>decomposable</u> if $V = D(V)$;

(d) <u>absolutely order-convex</u> if $S(V) \subseteq V$;

(e) <u>absolutely dominated</u> if $V \subseteq S(V)$;

(f) <u>solid</u> if $V = S(V)$;

(g) <u>positively order-convex</u> if $(V - C) \cap C \subseteq V$;

(h) <u>positively dominated</u> if $V \subseteq V \cap C - C$.

If V is order-convex and symmetric then $S(V) \subseteq V$ and $S(V)$ is the largest solid set contained in V ; in this case, $S(V)$ is referred to as the <u>solid kernel</u> of V . If V is positively order-convex, then

$$V \cap C = D(V) \cap C = S(V) \cap C = B(V) \cap C .$$

Let (E, C) and (F, K) be ordered vector spaces which form an ordered duality on the right, and let M be a subset of F . Since F can be regarded as a subspace of E^* and since K can be regarded as a subcone of the dual cone C^* . We need to distinguish the order-convex hull of M with respect to K as well as C^* , therefore we define

$$F(M) = (M + K) \cap (M - K) ;$$

$$F^*(M) = (M + C^*) \cap (M - C^*) ;$$

$$S(M) = \{y \in F : \exists\, w \in M \cap K \text{ such that } w \pm y \in K\} ;$$

$$S^*(M) = \{f \in E^* : \exists\, g \in M \cap C^* \text{ such that } g \pm f \in C^*\} ;$$

$$B(M) = \{y \in F : \exists\, w, v \in M \cap K \text{ such that } v - y, w + y \in K\} ;$$

$$B^*(M) = \{f \in E^* : \exists\, g, h \in M \cap C^* \text{ such that } g - f, h + f \in C^*\} .$$

It is clear that $F(M) = F^*(M) \cap F$, $S(M) = S^*(M) \cap F$ and that $B(M) = B^*(M) \cap F$ because of $K = -C^o$. If F is an order-convex subspace of (E^*, C^*) then $F^*(M) = F(M)$, $S^*(M) = S(M)$ and $B^*(M) = B(M)$.

(1.1.7) Theorem. *Let* (E, C) *and* (F, K) *be ordered vector spaces which form an ordered duality on the right, and suppose that* V *is a convex, circled* $\tau(E, F)$-*neighbourhood of* 0 . *Then we have*

(a) $(F(V))^{\circ} = D(V^{\circ})$;

(b) $(D(V))^{\circ} = F(V^{\circ})$.

In particular, the polar of an o-*convex* (*resp.* *decomposable*) $\tau(E, F)$-*neighbourhood of* 0 *is decomposable* (*resp.* o-*convex*).

Proof. Since $(V + C)$ and $(V - C)$ are convex $\tau(E, F)$-neighbourhoods of 0 , it follows from Lemma (1.1.4) that

$$(F(V))^{\circ} = \overline{co}\{(V + C)^{\circ} \cup (V - C)^{\circ}\}$$
$$= co\{(V + C)^{\circ} \cup (V - C)^{\circ}\} .$$

On applying Lemma (1.1.5) (a), we have

$$(F(V))^{\circ} = co\{-(V^{\circ} \cap K) \cup (V^{\circ} \cap K)\} = D(V^{\circ}) ,$$

as required in (a).

By the bipolar theorem and Lemma (1.1.5)(b), we have

$$(D(V))^{\circ} = (-(V \cap C))^{\circ} \cap (V \cap C)^{\circ} = (V^{\circ} + K) \cap (V^{\circ} - K) = F(V^{\circ}) ,$$

as required in (b).

(1.1.8) Theorem. *Let* (E, C) *and* (F, K) *be ordered vector spaces which form an ordered duality on the right, and suppose that* V *is a convex circled* $\tau(E, F)$-*neighbourhood of* 0 *in* E . *Then the following statements hold:*

(a) <u>If</u> C <u>is closed, then</u>

$$(F(V^o))^o = \overline{D(\overline{V})} .$$

(b) $$(D(V^o))^o = \overline{F(\overline{V})} . \tag{1.8}$$

<u>Moreover, if</u> V^o <u>is decomposable</u> (<u>in particular</u>, V <u>is order-convex</u>), <u>then the closure</u> $\overline{V}$ <u>of</u> V <u>is order-convex</u>, i.e.,

$$\overline{V} = F(\overline{V}) . \tag{1.9}$$

<u>Proof</u>. (a) As $V^o + K$ and $V^o - K$ are $\sigma(F, E)$-closed absolutely convex subsets of F, it follows from Lemma (1.1.4) that

$$(F(V^o))^o = \overline{co}((V^o + K)^o \cup (V^o - K)^o) .$$

In view of the bipolar theorem and the closedness of C, we have, by Lemma (1.1.5) (a) that

$$(F(V^o))^o = \overline{co}(-(\overline{V} \cap C) \cup (\overline{V} \cap C)) = \overline{D(\overline{V})} .$$

(b) According to the bipolar theorem and Theorem (1.1.7)(a), we have

$$(F(\overline{V}))^o = D(V^o) \subset (D(V^o))^{oo} ,$$

and hence $(D(V^o))^o \subseteq \overline{F(\overline{V})}$.

Conversely, let $x \in F(\overline{V})$ and let $y, z,$ in $\overline{V}$, be such that

$$y \leq x \leq z .$$

We first show that

$$|g(x)| \leq 1 \quad \text{for all} \quad g \in V^o \cap K .$$

In fact, as $y, z \in \overline{V}$ and $\overline{V}$ is circled, it follows that $|g(y)| \leq 1$ and $|g(z)| \leq 1$. Since $g \in K$, we have

$$-1 \leq g(y) \leq g(x) \leq g(z) \leq 1$$

which obtains our required assertion.

Now for any $f \in D(V^o)$, there exist $g, h \in V^o \cap K$ and $\lambda \in [0, 1]$ such that $f = \lambda g - (1 - \lambda)h$, hence

$$|f(x)| \leq \lambda |g(x)| + (1 - \lambda)|h(x)| \leq 1$$

which shows that $x \in (D(V^o))^o$, consequently $F(\overline{V}) \subseteq (D(V^o))^o$. The closedness of $(D(V^o))^o$ implies that $\overline{F(\overline{V})} \subseteq (D(V^o))^o$. Therefore we obtain the equality (1.8).

Finally, if V is order-convex, then V^o is decomposable by Theorem (1.1.7), therefore we complete the proof by showing that if V^o is decomposable, then $\overline{V} = F(\overline{V})$.

In fact, as $V^o = D(V^o)$, it follows from Formula (1.8) that

$$\overline{V} = V^{oo} = (D(V^o))^o = \overline{F(\overline{V})} \supseteq F(\overline{V}) ,$$

and hence that $\overline{V} = F(\overline{V})$ because $\overline{V}$ is always contained in $F(\overline{V})$.

It is worthwhile to notice from (1.9) that the order-convex hull of a closed, circled o-convex $\tau(E, F)$-neighbourhood of o is <u>closed</u>.

As another application of Bonsall's theorem, we establish the second duality theorem as follows.

(1.1.9) Theorem (Jameson). <u>Let</u> (E, C) <u>and</u> (F, K) <u>be ordered vector spaces which form an ordered duality on the right</u>, <u>and suppose that</u> V *is a convex* $\tau(E, F)$*-neighbourhood of* 0 *in* E *. Then we have*

$$(S(V))^o = S(V^o) .$$

Consequently, if V *is absolutely dominated (resp. absolutely order-convex, solid) then* V^o *is absolutely order-convex (resp. absolutely dominated, solid).*

Proof. We first show that $S(V^o) \subseteq (S(V))^o$. Let $f \in S(V^o)$ and let g , in $V^o \cap K$, be such that $\pm f \leq g$. For any $x \in S(V)$ there exists $v \in V \cap C$ such that $\pm x \leq v$. Then we have

$$0 \leq \langle v+x , g-f \rangle = \langle v , g \rangle - \langle v , f \rangle + \langle x, g \rangle - \langle x , f \rangle$$

$$0 \leq \langle v-x , g+f \rangle = \langle v , g \rangle + \langle v , f \rangle - \langle x , g \rangle - \langle x , f \rangle .$$

Summing up, it follows that

$$\langle x , f \rangle \leq \langle v , g \rangle \leq 1$$

hence $f \in (S(V))^o$.

To prove the inclusion $(S(V))^o \subseteq S(V^o)$, let $f \in (S(V))^o$ and define

$$q(v) = \sup \{\langle x , f \rangle : \pm x \leq v\} \quad (v \in C) .$$

Then q is a superlinear functional on C such that

$$q(v) \leq p(v) \quad \text{for all} \quad v \in C ,$$

where p is the gauge of V , because of

$$\pm x \leq v \in V \Rightarrow \langle x , f \rangle \leq 1 .$$

By Bonsall's theorem, there exists $g \in E^*$ such that

$$q(v) \leq g(v) \quad (v \in C)$$

and

$$g(x) \leq p(x) \quad (x \in E) .$$

Since V is a $\tau(E, F)$-neighbourhood of 0, it follows that $g \in V^{o}$. As

$$\langle \pm v, f\rangle \leq q(v) \leq g(v) \quad \text{for all} \quad v \in C .$$

We see that $\pm f \leq g$, thus $f \in S(V^{o})$.

We now present a dual result of Theorem (1.1.9), which is essential due to Ng and Duhoux [1], as follows

(1.1.10) Theorem. Let (E, C) and (F, K) be ordered vector spaces which form an ordered duality on the right, and suppose that V is an open convex circled $\tau(E, F)$-neighbourhood of 0 in E. Then the following statements hold.

(a) If V^{o} is absolutely dominated, then $\overline{V}$ is absolutely order-convex.

(b) If $D(V)$ is a $\tau(E, F)$-neighbourhood of 0 (and surely C is generating), and if V^{o} is absolutely order-convex, then V is absolutely dominated.

Proof. The proof of (a) is similar to that of the first part of Theorem (1.1.9). To prove (b), let $x_o \in V$ and $\delta = p(x_o)$, where p is the gauge of V. Suppose on the contrary that there does not exist $v \in V \cap C$ such that $\pm x_o \leq v$, then we are going to seek a contradition. As V is open, $p(x_o) = \delta < 1$, there exists $\varepsilon > 0$ such that $\delta + 3\varepsilon < 1$. Let $U = (\delta + \varepsilon)V$. Then $x_o \in U$ and U^{o} is an absolutely order-convex subset of (F, K). Let $W = D(\varepsilon V)$. By the hypothesis, W is a $\tau(E, F)$-neighbourhood of 0. Consider the product space $E \times E$ ordered by the product cone $C \times C$, and let

$$H = \{(x, y) \in E \times E : \exists\, u \in U \text{ such that } x \leq u \text{ and } y \leq u\}$$

Then H is a convex set containing $-C \times (-C)$ and (u, u) for all $u \in U$. We now claim that $(x_0 + W) \times (-x_0 + W)$ and H are disjoint. In fact, if not, there exist $w, w' \in W$ and $u_0 \in U$ such that

$$x_0 + w \leq u_0 \quad \text{and} \quad -x_0 + w' \leq u_0 .$$

Since $W = D(\varepsilon V)$, we can assume that

$$w = \varepsilon[\lambda v_1 - (1 - \lambda)v_2] \quad \text{and} \quad w' = \varepsilon[\mu v_1' - (1 - \mu)v_2'] ,$$

where $\lambda, \mu \in [0, 1]$, v_j and v_j' are in $V \cap C$ $(j = 1, 2)$. Let

$$v = u_0 + \varepsilon[(1 - \lambda)v_2 + (1 - \mu)v_2'] .$$

Then $p(v) \leq (\delta + \varepsilon) + 2\varepsilon < 1$, and thus $v \in V \cap C$ and $\pm x_0 \leq v$, contrary to our assumption on x_0. As $W \times W$ is a $\tau(E, F) \times \tau(E, F)$-neighbourhood of 0 in $E \times E$, it follows that $(x_0, -x_0)$ does not belong to the $\tau(E, F) \times \tau(E, F)$-closure of H.

By the strong separation theorem, there exists a $\tau(E, F) \times \tau(E, F)$-continuous linear functional ϕ on $E \times E$ such that

$$0 \leq \sup \phi(H) < 1 < \phi(x_0, -x_0) .$$

Since H contains $-C \times (-C)$, it follows that $\phi \geq 0$ on $C \times C$. Let us now define, for each $x \in E$, that

$$g_1(x) = \phi(x, 0) \quad \text{and} \quad g_2(x) = \phi(0, x) .$$

Then g_1 and g_2 are $\tau(E, F)$-continuous linear functional on E, and hence $g_1, g_2 \in F$. On the other hand, setting

$$f = g_1 + g_2 \quad \text{and} \quad h = g_1 - g_2 .$$

Then $\pm h \leq f$ and

$$f(x) = \phi(x, x) < 1 \quad \text{for all} \quad x \in U$$

because of $\{(x, x) : x \in U\} \subset H$, hence $f \in U^o$ and thus $h \in U^o$ in view of the absolute order-convexity of U^o; in particular,

$$h(x_o) \leq 1$$

which contradicts the fact that $h(x_o) = \phi(x_o, -x_o) > 1$.

The contradition shows that there must exist $v \in V \cap C$ such that $\pm x_o \leq v$, therefore V is absolutely dominated.

(1.1.11) Proposition. _Let_ (E, C) _and_ (F, K) _be ordered vector spaces which form an ordered duality on the right, and suppose that_ V _is a convex circled_ $\tau(E, F)$_-neighbourhood of_ 0 _in_ E. _Then the following statements hold:_

(a) _If_ $F(V)$ _is a_ $\tau(E, F)$_-neighbourhood of_ 0, _then_

$$(D(F(V)))^o = F(D(V^o)) = B(V^o) .$$

(b) _If_ $S(V)$ _is a_ $\tau(E,F)$_-neighbourhood of_ 0 _(hence_ $C - C = E$), _then_

$$(D(F(V)))^o = F(S(V^o)) .$$

(c) _If_ $D(V)$ _is a_ $\tau(E, F)$_-neighbourhood of_ 0 _(hence_ $C - C = E$), _then_

$$(B(V))^o = D(F(V^o)) \quad \text{and}$$
$$(B(V))^o \cap K = F(V^o) \cap K = S(V^o) \cap K$$
$$= \{f \in K : f \leq g \ \text{for some} \ g \in V^o \cap K\} .$$

<u>Proof</u>. (a) As $F(V)$ is a $\tau(E, F)$-neighbourhood of 0, we have, in view of Theorem (1.1.7) and Lemma (1.1.6), that

$$(D(F(V)))^{\circ} = F(D(V^{\circ})) = B(V^{\circ}) .$$

(b) The order-convexity of $F(V)$ insures that $F(V) \cap C = S(V) \cap C$. Since $S(V)$ is a $\tau(E, F)$-neighbourhood of 0, we have, in view of Lemma (1.1.5) (b) and Theorem (1.1.9), that

$$\begin{aligned}(D(F(V)))^{\circ} &= (-(F(V) \cap C))^{\circ} \cap (F(V) \cap C)^{\circ} = (-(S(V) \cap C))^{\circ} \cap (S(V) \cap C)^{\circ} \\ &= (S(V^{\circ}) + K) \cap (S(V^{\circ}) - K) = F(S(V^{\circ})) .\end{aligned}$$

(c) Since $D(V)$ is a $\tau(E, F)$-neighbourhood of 0, it follows from Theorem (1.1.7) and Lemma (1.1.6) that

$$(B(V))^{\circ} = (F(D(V)))^{\circ} = D(F(V^{\circ})) .$$

On the other hand, it is easily seen that

$$S(V^{\circ}) \cap K = \{f \in K : f \leq g \text{ for some } g \in V^{\circ} \cap K\} = F(V^{\circ}) \cap K .$$

Since $D(F(V^{\circ})) \cap K = F(V^{\circ}) \cap K$, we conclude that

$$(B(V))^{\circ} \cap K = F(V^{\circ}) \cap K$$

which obtains the required equalities.

For the sake of completeness we include the following result concerning the duality problem for positively order-convex sets and positively dominated sets, which can be verified in the same way as Theorems (1.1.9) and (1.1.10).

(1.1.12) Theorem. <u>Let</u> (E, C) <u>and</u> (F, K) <u>be ordered vector spaces which form an ordered duality on the right</u>, <u>and suppose that</u> V <u>is an open convex circled</u> $\tau(E, F)$-<u>neighbourhood of</u> 0 <u>in</u> E. <u>Then the following statements hold</u>.

(a) *If* V *is positively order-convex, then* V° *is positively dominated. Dually, if* V° *is positively dominated, then* $\overline{V}$ *is positively order-convex.*

(b) *If* $D(V)$ *is a* $\tau(E, F)$*-neighbourhood of* 0 *in* E *(and surely* C *is generating), then* V *is positively dominated if and only if* V° *is positively order-convex.*

1.2. Seminorms on ordered vector spaces

A seminorm p defined on an ordered vector space (E, C) is said to be

(1) *strongly monotone* if

$$y \leq x \leq z \Rightarrow p(x) \leq \max\,(p(y), p(z)) \;;$$

(2) *absolutely monotone* if

$$-u \leq x \leq u \Rightarrow p(x) \leq p(u) \;;$$

(3) *monotone* if

$$0 \leq u \leq w \Rightarrow p(u) \leq p(w) \;.$$

For a seminorm p on (E, C), we have the following trivial implications:

strongly monotone $\Rightarrow$ absolutely monotone $\Rightarrow$ monotone .

Suppose now that p is a seminorm on (E, C) . We set throughout this section that

$$V = \{x \in E : p(x) < 1\} \quad \text{and} \quad V_1 = \{x \in E : p(x) \leq 1\} \;.$$

Clearly V is a convex circled $\tau(E, E^*)$-neighbourhood of 0 , thus p is the gauge of V as well as of V_1 , consequently we have

$$p(x) = \sup\,\{|h(x)| : h \in V^{\pi}\} \qquad (x \in E) \tag{2.1}$$

(1.2.1) Lemma. (a) p *is strongly monotone if and only if* V *is order-convex. In this case we have*

$$p(x) = \sup\{|f(x)| : f \in V^{\pi} \cap C^*\} \qquad (x \in E) .$$

(b) p <u>is absolutely monotone</u> $\Leftrightarrow$ V <u>is absolutely order-convex</u> $\Leftrightarrow$

$$p(x) \leq \inf\{p(u) : u \pm x \in C\} \quad (x \in E) . \tag{2.2}$$

(c) p <u>is monotone</u> $\Leftrightarrow$ V <u>is positively order-convex</u> $\Leftrightarrow$

$$p(u) = \sup\{f(u) : f \in V^{\pi} \cap C^*\} \qquad (u \in C) . \tag{2.3}$$

<u>Proof</u>. (a) The necessity is trivial. For the sufficiency, let $y \leq x \leq z$ and $\alpha = \max(p(y), p(z))$. For any $\varepsilon > 0$, $(\alpha + \varepsilon)^{-1}y$ and $(\alpha + \varepsilon)^{-1}z$ belong to V, hence $(\alpha + \varepsilon)^{-1}x \in V$, consequently $p(x) < \alpha + \varepsilon$. This implies that $p(x) \leq \max(p(y), p(z))$ which shows that p is strongly monotone.

In view of (2.1), it is clear that

$$\sup\{|f(x)| : f \in V^{\pi} \cap C^*\} \leq p(x) .$$

If now p is strongly monotone, then by Theorem (1.1.7)(a), V^{π} is decomposable, thus any $h \in V^{\pi}$ can be expressed as the form $h = \lambda f - (1 - \lambda)g$ for some $\lambda \in [0, 1]$ and $f, g \in V^{\pi} \cap C^*$. From this we obtain

$$|h(x)| \leq \lambda |f(x)| + (1 - \lambda)|g(x)| \leq \sup\{|\phi(x)| : \phi \in V^{\pi} \cap C^*\} ,$$

consequently

$$p(x) \leq \sup\{|\phi(x)| : \phi \in V^{\pi} \cap C^*\} .$$

(b) The proof is similar to that given in (a).

(c) If p is monotone, then V is clearly positively order-convex. Suppose now that V is positively order-convex, then by Theorem (1.1.12)(a), V^{π} is positively dominated. For any $h \in V^{\pi}$, there exists $f \in V^{\pi} \cap C^*$

such that $h \leq f$, thus

$$h(u) \leq f(u) \leq \sup \{\phi(u) : \phi \in V^\pi \cap C^*\}.$$

As V^π is symmetric, we have $|h(u)| \leq \sup \{\phi(u) : \phi \in V^\pi \cap C^*\}$, and hence

$$p(u) \leq \sup \{\phi(u) : \phi \in V^\pi \cap C^*\} .$$

Finally, if the inequality (2.3) holds and if $0 \leq u \leq w$, then

$$0 \leq f(u) \leq f(w) \quad \text{for all} \quad f \in V^\pi \cap C^* ,$$

consequently $p(u) \leq p(w)$. Therefore p is monotone.

(1.2.2) Lemma. V <u>is decomposable if and only if</u>

$$p(x) = \inf \{p(u) + p(w) : u, w \in C \text{ and } u - w = x\} \quad (x \in E) \qquad (2.4)$$

<u>Proof</u>. <u>Necessity</u>. Clearly

$$p(x) \leq \inf \{p(u) + p(w) : u, w \in C \text{ and } u - w = x\} .$$

For any $\varepsilon > 0$, $(p(x) + \varepsilon)^{-1}x \in V$, hence there exist $\lambda \in [0, 1]$ and $x_1, x_2 \in V \cap C$ such that $(p(x) + \varepsilon)^{-1}x = \lambda x_1 - (1 - \lambda)x_2$. Letting

$$u = \lambda(p(x) + \varepsilon)x_1 \quad \text{and} \quad w = (1 - \lambda)\ (p(x) + \varepsilon)x_2 .$$

Then $u, w \in C$ are such that $u - w = x$, and

$$p(u) + p(w) = (p(x) + \varepsilon)(\lambda p(x_1) + (1 - \lambda)p(x_2)) < p(x) + \varepsilon .$$

Therefore we get the required equality (2.4).

<u>Sufficiency</u>. Let $x \in V$. Then $p(x) < 1$, hence there exist $u, w \in C$ such that

$$u - w = x \quad \text{and} \quad p(u) + p(w) < 1 .$$

Take positive real numbers λ_1 and λ_2 such that

$$0 \leqslant p(u) < \lambda_1 \ , \ 0 \leqslant p(w) < \lambda_2 \quad \text{and} \quad \lambda_1 + \lambda_2 = 1 \ .$$

Then u/λ_1 and w/λ_2 are in $V \cap C$ and $x = \lambda_1(u/\lambda_1) - \lambda_2(w/\lambda_2) \in D(V)$. Therefore V is decomposable.

(1.2.3) Lemma. (a) V <u>is absolutely dominated if and only if for any</u> $x \in E$ <u>and</u> $\varepsilon > 0$, <u>there exists</u> $u \in C$ <u>such that</u>

$$\pm x \leqslant u \quad \text{and} \quad p(u) < p(x) + \varepsilon \ .$$

(b) V <u>is positively dominated if and only if for any</u> $x \in E$ <u>and</u> $\varepsilon > 0$, <u>there exist</u> $u, w \in C$ <u>such that</u>

$$-u \leqslant x \leqslant w \quad \text{and} \quad \max(p(u), p(w)) < p(x) + \varepsilon \ .$$

The proof is similar to that given in Lemma (1.2.2) and therefore will be omitted.

(1.2.4) Corollary. (a) V <u>is solid if and only if</u>

$$p(x) = \inf \{p(u) : u \in C \ \ \underline{\text{and}} \ \ u \pm x \in C\} \ (x \in E). \qquad (2.5)$$

(b) <u>If</u> V <u>is order-convex and positively dominated, then</u>

$$p(x) = \inf \{\max(p(u), p(w)) : u, w \in C \ \ \underline{\text{and}} \ \ -u \leqslant x \leqslant w\} \ . \qquad (2.6)$$

<u>Proof</u>. (a) The sufficiency follows from Lemma (1.2.3) (a) and (1.2.1) (b) . For the necessity, by Lemma (1.2.1) (b), we have

$$p(x) \leqslant \inf \{p(u) : u \in C \ \text{and} \ u \pm x \in C\} \ ,$$

and by Lemma (1.2.3) (a), we have for any $\varepsilon > 0$ that there exists $u \in C$ such that

$$\pm x \leq u \quad \text{and} \quad p(u) < p(x) + \varepsilon .$$

Therefore the equality (2.5) holds.

(b) Follows from Lemma (1.2.1) (a) and (1.2.3) (b) .

A seminorm p on (E, C) is called a <u>Riesz seminorm</u> (or <u>regular seminorm</u>) if

$$p(x) = \inf \{p(u) : u \in C \text{ and } u \pm x \in C\} \quad (x \in E) .$$

By making use of Corollary (1.2.4) (a), it is easily seen that p is regular if and only if V is solid.

Let p be a seminorm on (E, C) and

$$V = \{x \in E : p(x) < 1\} .$$

Then $D(V) \subseteq S(V) \subseteq B(V)$ and $B(V)$ is o-convex, circled. If C is generating, then $D(V)$, and surely $S(V)$ and $B(V)$, is absorbing; denote by p_D , p_S and p_B are gauges of $D(V)$, $S(V)$ and $B(V)$ respectively, we have the following result.

(1.2.5) Lemma. <u>Let</u> (E, C) <u>be an ordered vector space, let</u> C <u>be generating and let</u> p <u>be a seminorm on</u> E . <u>Then we have</u>

$$p_D(x) = \inf \{p(x_1) + p(x_2) : x_1 , x_2 \in C \text{ and } x_1 - x_2 = x\} \quad (2.7)$$

$$p_S(x) = \inf \{p(u) : u \pm x \in C\} \quad (2.8)$$

$$p_B(x) = \inf \{\max (p(y), p(z)) : y, z \in C \text{ and } -y \leq x \leq z\} \quad (2.9)$$

whenever $x \in E$, *and* $p_D(u) = p(u)$ *for all* $u \in C$.

Furthermore, if p *is monotone, then*

$$p_B(u) = p_S(u) = p_D(u) \qquad \textit{for all} \quad u \in C . \tag{2.10}$$

Proof. Let $\mu = \inf \{p(x_1) + p(x_2) : x_1, x_2 \in C \text{ and } x_1 - x_2 = x\}$. For any $\alpha > 0$ with $x \in \alpha D(V)$, we have $x = \alpha\lambda u - \alpha(1 - \lambda)w$ for some $\lambda \in [0, 1]$ and $u, w \in V \cap C$, thus $p(\alpha\lambda u) + p(\alpha(1 - \lambda)w) < \alpha$. From this it follows that $\mu \leq p_D(x)$.

On the other hand, for any $\varepsilon > 0$ there exist $x_1, x_2 \in V \cap C$ such that

$$x_1 - x_2 = x \quad \text{and} \quad p(x_1) + p(x_2) < \mu + \varepsilon .$$

Letting $k_i = p(x_i) + \frac{\varepsilon}{2}$ $(i = 1, 2)$. Then $k_i^{-1} x_i \in V \cap C$ $(i = 1, 2)$ and

$$x = (k_1 + k_2) \left(\frac{k_1}{k_1 + k_2} \frac{x_1}{k_1} - \frac{k_2}{k_1 + k_2} \frac{x_2}{k_2} \right) \in (k_1 + k_2) D(V) .$$

It follows that

$$p_D(x) \leq k_1 + k_2 < \mu + 2\varepsilon ,$$

therefore we obtain the equality (2.7) .

By a similar argument we can prove the equalities (2.8) and (2.9) .

Note that $D(V) \subseteq V$; we have $p(x) \leq p_D(x)$ $(x \in E)$. For any $u \in C$, as $p_D(u) \leq p(u) + p(0) = p(u)$, we obtain $p_D(u) = p(u)$ $(u \in C)$.

To verify the equality (2.10) , it is sufficient to show that

$$p(u) \leq p_B(u) \qquad \text{for all} \quad u \in C .$$

In fact, for any $\varepsilon > 0$, there exist $y, z \in C$ such that

$$-y \leq u \leq z \quad \text{and} \quad \max(p(y), p(z)) < p_B(u) + \varepsilon .$$

As $u \in C$ and p in monotone, we have

$$p(u) \leq p(z) \leq \max(p(y), p(z)) < p_B(u) + \varepsilon ,$$

and thus $p(u) \leq p_B(u)$.

Let (E, C) be an ordered vector space and G a subspace of E. Then $G \cap C$ is always a subcone of C, and G is decomposable if and only if $G = G \cap C - G \cap C$. In particular, if K is a subcone of C and if $G = K - K$, then G is decomposable. In the remainder of this section, G is assumed to be a decomposable subspace of E and $K = G \cap C$. Then K is a generating cone in G. If λ is a positive sublinear functional defined on K, then we can define a functional p_λ on G by setting

$$p_\lambda(x) = \inf \{\lambda(x_1) + \lambda(x_2) : x_1, x_2 \in K, x_1 - x_2 = x\} \quad (x \in G) \tag{2.11}$$

It is not hard to show that p_λ is a seminorm on G such that

$$p_\lambda(u) \leq \lambda(u) \qquad \text{for all} \quad u \in K . \tag{2.12}$$

Furthermore, we have

(1.2.6) Lemma. Let G be a decomposable subspace of (E, C), let $K = G \cap C$ and let λ be a monotone sublinear functional on K. Define p_λ on G by setting

$$p_\lambda(x) = \inf \{\lambda(x_1) + \lambda(x_2) : x_1, x_2 \in K, x_1 - x_2 = x\} \quad (x \in G).$$

Suppose further that

$$\Sigma_\lambda^+ = \{u \in K : \lambda(u) < 1\} \quad \text{and} \quad W_\lambda = \{x \in G : p_\lambda(x) < 1\} .$$

Then the following assertions hold:

(1) $p_\lambda(u) = \lambda(u) \quad (u \in K)$, and hence $\Sigma_\lambda^+ = W_\lambda \cap K$.

(2) $p_\lambda = p_{\lambda D}$ and $W_\lambda = D(W_\lambda) = co((-\Sigma_\lambda^+) \cup \Sigma_\lambda^+)$.

(3) The gauge of $S(W_\lambda) = \cup\{[-u, u] : u \in \Sigma_\lambda^+\}$, denoted by $p_{\lambda S}$, is

$$p_{\lambda S}(x) = \inf\{\lambda(u) : u \pm x \in K\} . \quad (x \in G) .$$

(4) The gauge of $B(W_\lambda) = \cup\{[-v, u] : v, u \in \Sigma_\lambda^+\}$, denoted by $p_{\lambda B}$, is

$$p_{\lambda B}(x) = \inf\{\max(\lambda(v), \lambda(u)) : v,u \in K, -v\leq x\leq u\}. \quad (x \in G).$$

Proof. The conclusion (2), (3) and (4) follow from (1) by making use of (1.2.5). To prove $p_\lambda = \lambda$ on K , it is sufficient to verify that $\lambda(u) \leq p_\lambda(u) \quad (u \in K)$. In fact, for any $u \in K$ and any $\delta > 0$, there exist $x_1, x_2 \in K$ such that

$$u = x_1 - x_2 \quad \text{and} \quad \lambda(x_1) + \lambda(x_2) < p_\lambda(u) + \delta .$$

Since $x_1 + x_2 - u \in K$ and since λ is monotone, it follows that

$$\lambda(u) \leq \lambda(x_1 + x_2) \leq \lambda(x_1) + \lambda(x_2) < p_\lambda(u) + \delta .$$

As δ was arbitrary, we obtain the required inequality.

Finally, we would like to point out that if G, K and λ are assumed as in (1.2.6), and if we define

$$q_\lambda(x) = \inf\{\lambda(u) : u \pm x \in K\} \quad (x \in G) ,$$

$$V_\lambda = \{x \in G : q_\lambda(x) < 1\} ,$$

then q_λ is a seminorm on G such that

$$q_\lambda(u) = \lambda(u) \quad \text{for all } u \in K .$$

Consequently, q_λ is the gauge of $S(V_\lambda) = \cup\{[-u, u] : u \in V_\lambda \cap K\}$, that is $q_\lambda = q_{\lambda S}$, and

$$q_{\lambda D}(x) = \inf\{\lambda(x_1) + \lambda(x_2) : x_1, x_2 \in K , x_1 - x_2 = x\} = p_\lambda(x)$$

$$q_{\lambda B}(x) = \inf\{\max(\lambda(v),\lambda(u)) : v,u \in K , -v \leq x \leq u\} = p_{\lambda B}(x)$$

whenever $x \in G$.

Similarly, if we define

$$r_\lambda(x) = \inf\{\max(\lambda(v), \lambda(u)) : v, u \in K, -v \leq x \leq u\} \quad (x \in G)$$

$$U_\lambda \quad = \{x \in G : r_\lambda(x) < 1\} ,$$

then r_λ is a seminorm on G such that

$$r_\lambda(u) = \lambda(u) \quad \text{for all} \quad u \in K .$$

Consequently, r_λ is the gauge of $B(U_\lambda) = \cup\{[-u, u] : u \in U_\lambda \cap K\}$, that is $r_\lambda = r_{\lambda B}$, hence

$$r_{\lambda D} = p_\lambda \quad \text{and} \quad r_{\lambda S} = p_{\lambda S} .$$

1.3 Topologies on ordered vector spaces

An ordered convex space $(E, C, \mathcal{P})$ is said to be locally decomposable (resp. locally o-convex, locally solid) if $\mathcal{P}$ admits a neighbourhood base at 0 consisting of absolutely convex and decomposable (resp. o-convex and circled, convex and solid) sets in E. As Theorem (1.1.8)(b) shown, the closure of an o-convex circled $\mathcal{P}$-neighbourhood of 0 is order-convex, it follows that the locally o-convex topology $\mathcal{P}$ on E admits a neighbourhood base at 0 consisting closed o-convex circled sets in E. Clearly $(E, C, \mathcal{P})$ is a locally solid space if and only if it is locally decomposable and locally o-convex. Other characterizations of locally decomposable and locally solid spaces can be found in Wong and Ng [1], while characterizations of locally o-convex spaces can be found in Schaefer [1], Peressini [1] and Jameson [1].

Let $(E, C, \mathcal{P})$ be an ordered convex space and $\mathcal{U}$ a neighbourhood base at 0 for $\mathcal{P}$ consisting of convex circled sets in E. Setting

$$F(\mathcal{U}) = \{F(V) : V \in \mathcal{U}\}$$
$$D(\mathcal{U}) = \{D(V) : V \in \mathcal{U}\}$$
$$S(\mathcal{U}) = \{S(V) : V \in \mathcal{U}\} .$$

It is easily seen that $F(\mathcal{U})$ determines a unique locally o-convex topology, denoted by $\mathcal{P}_F$; moreover $\mathcal{P}_F$ is the least upper bound of all locally o-convex topologies which are coarser than $\mathcal{P}$. This topology $\mathcal{P}_F$ is referred to as the <u>locally o-convex topology associated with</u> $\mathcal{P}$. If C is generating, then $D(\mathcal{U})$(resp. $S(\mathcal{U})$) determines a unique locally decomposable (resp. locally solid) topology, denoted by $\mathcal{P}_D$ (resp. $\mathcal{P}_S$) ; moreover, $\mathcal{P}_D$ is the greatest lower bound of all locally decomposable topologies which are finer than $\mathcal{P}_D$. If, in addition, $\mathcal{P}$ is locally o-convex, then each $V \in \mathcal{U}$ can be chosen as o-convex, hence $S(V)$ is the largest solid set in E contained in V ; consequently $\mathcal{P}_S$ is the greatest lower bound of all locally solid topologies which are finer than $\mathcal{P}$. The topology $\mathcal{P}_D$ (resp. $\mathcal{P}_S$) is called the <u>locally decomposable</u> (resp. <u>locally solid</u>) topology <u>associated with</u> $\mathcal{P}$. The construction of $\mathcal{P}_F$ is essentially due to Namioka [1] and that of $\mathcal{P}_D$ to Wong and Cheung [1]. Recently Walsh [1] also give the construction of $\mathcal{P}_D$. For further information about $\mathcal{P}_F$ and $\mathcal{P}_D$, we refer the reader to Wong and Ng [1].

It is easily seen from Theorems (1.1.7) and (1.1.9) that the topological dual E' of a locally decomposable (resp. locally o-convex, locally solid) space is an order-convex (resp. decomposable, solid) subspace of (E^*, C^*). Therefore if $(E, C, \mathscr{P})$ is an ordered convex space whose topological dual E' is order-convex, and if $E = C-C$, then $\mathscr{P}_D$ is consistent with $\langle E, E'\rangle$; in particular, if the order-convex hull <u>in</u> (E^*, C^*) of each $\mathscr{P}$-equicontinuous subset of E' is $\mathscr{P}$-equicontinuous, then E' is order-convex in (E^*, C^*). Hence we obtain, in view of Theorems (1.1.7) and (1.1.8), the following dual criterion for the topology of $\mathscr{P}_F$ and $\mathscr{P}_D$.

(1.3.1) Theorem. <u>For an ordered convex space</u> $(E, C, \mathscr{P})$, <u>the following statements hold</u>:

(a) $\mathscr{P}_F$ <u>is the topology on</u> E <u>of uniform convergence on the decomposable</u> $\mathscr{P}$-<u>equicontinuous subsets of</u> E'; <u>consequently</u> $\mathscr{P}$ <u>is locally o-convex</u> (<u>i.e.</u> $\mathscr{P} = \mathscr{P}_F$) <u>if and only if each</u> $\mathscr{P}$-<u>equicontinuous subset of</u> E' <u>is contained in a decomposable</u> $\mathscr{P}$-<u>equicontinuous subset of</u> E'.

(b) <u>If</u> $E = C - C$, <u>then</u> $\mathscr{P}_D$ <u>is the topology on</u> E <u>of uniform convergence on the order-convex hulls in</u> (E^*, C^*) <u>of all</u> $\mathscr{P}$-<u>equicontinuous subsets of</u> E', <u>consequently</u> $\mathscr{P}$ <u>is locally decomposable</u> (i.e., $\mathscr{P} = \mathscr{P}_D$) <u>if and only if the order-convex hull in</u> (E^*, C^*) <u>of each</u> $\mathscr{P}$-<u>equicontinuous subset of</u> E' <u>is</u> $\mathscr{P}$-<u>equicontinuous</u>.

The preceding theorem has many important applications, we mention a few below.

(1.3.2) Corollary. <u>For any ordered convex space</u> $(E, C, \mathscr{P})$, <u>the following statements hold</u>.

(a) <u>The topological dual</u> $(E, \mathscr{P}_F)'$ <u>of</u> $(E, \mathscr{P}_F)$ <u>is the decomposable kernel of</u> $E' = (E, \mathscr{P})'$, <u>hence</u> $(E, \mathscr{P}_F)' = C' - C'$.

(b) <u>If</u> C <u>is generating, then the topological dual</u> $(E, \mathscr{P}_D)'$ <u>of</u> $(E, \mathscr{P}_D)$ <u>is the order-convex hull of</u> E' <u>in</u> (E^*, C^*) , <u>hence</u> $(E, \mathscr{P}_D)' = (E' + C^*) \cap (E' - C^*)$.

Since $E' = (E, \mathscr{P})'$ is total over E , it follows from a well-known result (see Schaefer [1 , p.125]) that E' is $\sigma(E_D', E)$-dense in E_D' , where $E_D' = (E, \mathscr{P}_D)' = (E' + C^*) \cap (E' - C^*)$. On the other hand, $\mathscr{P}_F$ need not be Hausdorff. But a well-known result shows that $\mathscr{P}_F$ is Hausdorff if and only if $C' - C'$ is total over E ; therefore, if the $\mathscr{P}$-closure of C is proper, then $C' - C'$ is $\sigma(E', E)$-dense in E' because the $\mathscr{P}$-closure of C coincides with the $\mathscr{P}_F$-closure of C .

(1.3.3) Corollary. <u>Let</u> $(E, C, \mathscr{P})$ <u>be an ordered convex space with the topological dual</u> E' . <u>Denote by</u> $\sigma(E, E')$ (<u>resp.</u> $\tau(E, E')$ <u>the weak topology</u> (<u>resp.</u> <u>Mackay topology</u>.) <u>Then the following statements hold</u>:

(a) $(E, C, \sigma(E, E'))$ <u>is a locally</u> o-<u>convex space if and only if</u> $E' = C' - C'$; <u>consequently if</u> $(E, C, \mathscr{P})$ <u>is locally</u> o-<u>convex, then so is</u> $(E, C, \sigma(E, E'))$.

(b) $(E, C, \tau(E, E'))$ <u>is locally decomposable if and only if</u>

$E = C - C$ <u>and</u> E' <u>is an order-convex subspace of</u> (E^*, C^*) ; <u>consequently if</u> $(E, C, \mathcal{P})$ <u>is locally decomposable, then so is</u> $(E, C, \tau(E, E'))$.

<u>Proof</u>. (a) The necessity is obvious. For the sufficiency, we know under the assumption that $\sigma_F(E, E')$ is consistent with the dual pair $\langle E, E'\rangle$, therefore $\sigma_F(E, E')$ coincides with $\sigma(E, E')$.

(b) The proof is similar to that given in (a) because $\tau_D(E, E')$ is consistent with $\langle E, E'\rangle$.

Let $L^1 = L^1[0, 1]$ be the ordered Banach space of all Lebesgue integrable real-valued functions on $[0, 1]$ equipped with the usual L^1 - norm $\|\cdot\|$ and the usual ordering. It is well known that the norm topology in L^1 is precisely the Mackay topology $\tau(L^1, L^\infty)$, and is strictly finer than $\sigma(L^1, L^\infty)$. Let V be the polar in L^1 of the constant function 1 (as an element in L^∞) and Σ the closed unit ball in L^1 . It is easily seen that

$$V \cap C - V \cap C \subseteq 2\Sigma .$$

Where C denotes the positive cone in L^1 . Since Σ <u>is not</u> a $\sigma(L^1, L^\infty)$-neighborhood of 0, $V \cap C - V \cap C$ must not be a neighborhood of 0. Therefore $(L^1, \sigma(L^1, L^\infty))$ is not a locally decomposable space, but it is well known that $(L^1, \tau(L^1, L^\infty))$ is a locally decomposable space.

(1.3.4) Proposition. <u>Let</u> $(E, C, \mathcal{P})$ <u>be a metrizable locally decomposable space</u>. <u>Denote by</u> $(\tilde{E}, \tilde{\mathcal{P}})$ <u>the completion of</u> $(E, \mathcal{P})$, <u>and by</u> $\bar{C}$ <u>the</u> $\tilde{\mathcal{P}}$<u>-closure of</u> C <u>in</u> $\tilde{E}$. <u>Then</u> $(\tilde{E}, \bar{C}, \tilde{\mathcal{P}})$ <u>is a metrizable locally decomposable space</u>. <u>In particular, if</u> K <u>is any cone in</u> $\tilde{E}$ <u>containing</u> $\bar{C}$,

<u>then</u> $(\tilde{E}, K, \tilde{\mathcal{P}})$ <u>is locally decomposable</u>.

<u>Proof</u>. Let upper-bars denote the $\tilde{\mathcal{P}}$-closures in $\tilde{E}$, let $G = \overline{C} - \overline{C}$ and let $\tilde{\mathcal{P}}_D$ be the locally decomposable topology on G associated with the relative topology on G induced by $\tilde{\mathcal{P}}$. Then $\tilde{\mathcal{P}}_D$ is metrizable bevause $\tilde{\mathcal{P}}$ is obviously metrizable, the cononical embedding map J from $(G, \tilde{\mathcal{P}}_D)$ into $(\tilde{E}, \tilde{\mathcal{P}})$ is continuous, and $(G, \tilde{\mathcal{P}}_D)$ is complete by Klee's theorem because $\overline{C}$ is $\tilde{\mathcal{P}}$-complete. Furthermore, we show that the canonical embedding map J is nearly open. Let V be a convex, circled $\mathcal{P}$-neighbourhood of 0 in E. Then $\overline{V}$ is a $\tilde{\mathcal{P}}$-neighbourhood of 0 in $\tilde{E}$, and $D(\overline{V})$ is a $\tilde{\mathcal{P}}_D$-neighbourhood of 0 in G. In order to show that J is nearly open, it is sufficient to verify that $\overline{D(\overline{V})} = \overline{J(D(\overline{V}))}$ is a $\tilde{\mathcal{P}}$-neighbourhood of 0 in E or, equivalently, $(D(\overline{V}))^{o}$ is a $\tilde{\mathcal{P}}$-equicontinuous subset of $(\tilde{E}, \tilde{\mathcal{P}})'$.

In fact, we first note that $E' = (\tilde{E}, \tilde{\mathcal{P}})'$. As $\overline{V}$ is a $\tilde{\mathcal{P}}$-neighbourhood of 0 in $\tilde{E}$, it follows from Theorem (1.1.7) that

$$(D(\overline{V}))^{o} = F(\overline{V}^{o}) = F(V^{o}) = (V^{o} + C') \cap (V^{o} - C'),$$

namely, $(D(\overline{V}))^{o}$ is the order-convex hull in (E', C') of V^{o}. As V^{o} is $\sigma(E', \tilde{E})$-bounded and as $(E', C', \sigma(E', \tilde{E}))$ is locally o-convex, it follows that $(D(V))^{o}$ is a $\sigma(E', \tilde{E})$-bounded subset of E'. On the other hand, since $(\tilde{E}, \tilde{\mathcal{P}})$ is a complete metrizable space and surely barrelled, $(D(\overline{V}))^{o}$ is a $\tilde{\mathcal{P}}$-equicontinuous subset of E'.

According to the open mapping theorem (see Schaefer [1, p.76]), J is an open mapping and hence J is a homeomorphism from $(G, \tilde{\mathcal{P}}_D)$ onto $(\tilde{E}, \tilde{\mathcal{P}})$. In particular, $\tilde{\mathcal{P}}_D = \tilde{\mathcal{P}}$ and $G = \tilde{E}$. Therefore $(\tilde{E}, \overline{C}, \tilde{\mathcal{P}})$ is a complete metrizable locally decomposable space.

(1.3.5) Theorem (Schaefer). *Let* $(E, C, \mathcal{P})$ *be an ordered convex space with the topological dual* E' *, and let* $\mathcal{C}$ *be a saturated family consisting of* $\sigma(E', E)$*-bounded subsets of* E' *for which the linear hull of* $\cup\{B, B \in \mathcal{C}\}$ *is* $\sigma(E', E)$*-dense in* E' *. Suppose further that* $\mathcal{P}$ *is the* $\mathcal{C}$*-topology on* E *. Then the following statements hold:*

(a) *If* C' *is an* $\mathcal{C}$*-cone, then* $\mathcal{P}$ *is a locally* o-*convex topology.*

(b) *Suppose that* $\mathcal{P}$ *is a locally o-convex topology. If* $\mathcal{P}$ *is consistent with the duality* $\langle E, E'\rangle$ *, then* C' *is a strict* $\mathcal{C}$*-cone* .

Proof. (a) Let V be a $\mathcal{P}$-neighbourhood of 0 in E . There exists a convex circled set A in $\mathcal{C}$ such that $A^{o} \subset V$, where A^{o} is the polar of A taken in E . Since C' is an $\mathcal{C}$-cone. A is contained in the $\sigma(E', E)$-closure $\overline{D(B)}$ for some convex circled set $B \in \mathcal{C}$. Then $(\overline{D(B)})^{o} = (D(B))^{o}$ is a $\mathcal{P}$-neighbourhood of 0 . Clearly $(D(B))^{o}$ is order-convex and contained in A^{o} and hence in V . This shows that $\mathcal{P}$ is locally o-convex.

(b) Since $\mathcal{P}$ is a consistent locally o-convex topology, it follows that $\mathcal{C}$ is a fundamental family of $\mathcal{P}$-equicontinuous sets and $E' = C' - C'$. By (1.3.1)(a), for any $A \in \mathcal{C}$ there exists $B \in \mathcal{C}$ such that $A \subseteq D(B)$, therefore C' is a strict $\mathcal{C}$-cone.

Krein-Grosberg's theorem says that an ordered normed space $(E, C, \|\cdot\|)$ is locally o-convex if and only if the dual cone C' is a strict $\mathcal{B}$-cone in $(E', \|\cdot\|)$; therefore Schaefer's theorem is a generalization of Krein-Grosberg's theorem.

(1.3.6) Corollary. <u>The completion of any locally o-convex space</u> $(E, C, \mathcal{P})$ <u>is a locally o-convex space</u>.

<u>Proof</u>. Denote by $(\tilde{E}, \tilde{\mathcal{P}})$ the completion of $(E, \mathcal{P})$, and by $\mathcal{U}$ the family of all o-convex circled, $\mathcal{P}$-neighbourhoods of 0, and by $\mathcal{U}^{\circ} = \{V^{\circ} : V \in \mathcal{U}\}$. Then C' is a strict $\mathcal{U}^{\circ}$-cone. If the upper-bars denote the $\tilde{\mathcal{P}}$-closures in $\tilde{E}$, then $\{\overline{V} : V \in \mathcal{U}\}$ is a neighbourhood base at 0 for $\tilde{\mathcal{P}}$, and $\tilde{\mathcal{P}}$ is the topology of uniform convergence on all V° $(V \in \mathcal{U})$, hence $(\tilde{E}, C, \tilde{\mathcal{P}})$ is a locally o-convex space by Schaefer's theorem.

Clearly, an ordered convex space $(E, C, \mathcal{P})$ is locally o-convex if and only if $(E, \overline{C}, \mathcal{P})$ is locally o-convex, where $\overline{C}$ is the $\mathcal{P}$-closure of C in E. Combining (1.3.4), we obtain the following.

(1.3.7) Corollary. <u>Let</u> $(E, C, \mathcal{P})$ <u>be a metrizable locally-solid space</u>. <u>Denote by</u> $(\tilde{E}, \tilde{\mathcal{P}})$ <u>the completion of</u> $(E, \mathcal{P})$, <u>and by</u> $\overline{C}$ <u>the</u> $\tilde{\mathcal{P}}$-<u>closure of</u> C <u>in</u> $\tilde{E}$. <u>Then</u> $(\tilde{E}, \overline{C}, \tilde{\mathcal{P}})$ <u>is a complete metrizable locally solid space</u>.

We are now going to consider the duality of local o-convexity and local decomposability. Let $(E, C, \mathcal{P})$ be an ordered topological vector space. C is called a <u>locally strict</u> $\mathcal{B}$-<u>cone</u> in $(E, \mathcal{P})$ if for any

$\mathcal{P}$-bounded set B in E and any $\mathcal{P}$-neighbourhood V of 0, there exists a subset A of E which is absorbed by V such that

$$B \subset A \cap C - A \cap C .$$

It is clear that any strict $\mathcal{B}$-cone in $(E, \mathcal{P})$ is a locally strict $\mathcal{B}$-cone in $(E, \mathcal{P})$, and that for an ordered normed space $(E, C, \|\cdot\|)$, C is a strict $\mathcal{B}$-cone in $(E, \|\cdot\|)$ if and only if it is a locally strict $\mathcal{B}$-cone in $(E, \|\cdot\|)$.

(1.3.8) Lemma. Let $(E, C, \mathcal{P})$ be an ordered convex space. If $(E, C, \mathcal{P})$ is locally decomposable then C is a locally strict $\mathcal{B}$-cone in $(E, \mathcal{P})$. The converse is true when $(E, \mathcal{P})$ is bornological.

Proof. Let $(E, C, \mathcal{P})$ be a locally decomposable space, let B be a $\mathcal{P}$-bounded set in E and let V be a convex, circled $\mathcal{P}$-neighbourhood of 0. Then $V \cap C - V \cap C$ is also a $\mathcal{P}$-neighborhood of 0. There exists $\lambda > 0$ such that $B \subset \lambda(V \cap C - V \cap C)$. Take $A = \lambda(V \cap C)$, then A is a positive subset of E, absorbed by V, and $B \subset A - A$, thus C is a locally strict $\mathcal{B}$-cone in $(E, \mathcal{P})$.

Conversely, suppose that E is bornological and that C is a locally strict $\mathcal{B}$-cone in $(E, \mathcal{P})$. Then we show that $(E, C, \mathcal{P})$ is locally decomposable. Let V be a convex, circled $\mathcal{P}$-neighborhood of 0. Then $W = V \cap C - V \cap C$ is a convex, circled subset of E. For any $\mathcal{P}$-bounded subset B of E, let A be a subset of E which is absorbed by V such that $B \subset A \cap C - A \cap C$. It is clear that $A \cap C - A \cap C$ is absorbed by W , consequently B is absorbed by W . As E is hornological, we conclude that W is a $\mathcal{P}$-neighbourhood of 0 and so $(E, C, \mathcal{P})$ is locally decomposable.

(1.3.9) Proposition. <u>Let</u> $(E, C, \mathcal{P})$ <u>be an infrabarrelled locally o-convex space</u>. <u>If</u> $(E', C', \beta(E', E))$ <u>is bornological then it is locally decomposable</u>. <u>In particular</u>, <u>for an ordered normed space</u> $(E, C, \|\cdot\|)$, <u>it is locally o-convex if and only if</u> $(E', C', \|\cdot\|)$ <u>is locally decomposable</u>.

<u>Proof</u>. According Schaefer's theorem, C' is a strict $\mathcal{B}$-cone in $(E', \beta(E', E))$, and hence the first assertion follows from (1.3.8) . The second assertion is also a consequence of Schaefer's theorem.

(1.3.10) Proposition. <u>Let</u> $(E, C, \mathcal{P})$ <u>be an infrabarrelled</u>, <u>ordered convex space with a countable fundamental system of</u> $\mathcal{P}$-<u>bounded subsets of</u> E . <u>Then</u> $(E, C, \mathcal{P})$ <u>is locally decomposable</u> <u>if and only if</u> $(E', C', \beta(E', E))$ <u>is locally o-convex and</u> E' <u>is order-convex in</u> E^* .

<u>Proof</u>. Suppose that $(E, C, \mathcal{P})$ is locally decomposable. Then E' is clearly order-convex in E^* . On the other hand, according to the assumption, $(E', C', \beta(E', E))$ is metrizable, therefore, it is sufficient to show that for any $\beta(E', E)$-bounded subset B of E' , the order-convex hull of B in E' is $\beta(E', E)$-bounded.

In fact, let B be a $\beta(E', E)$-bounded convex subset of E' . Then B is $\mathcal{P}$-equicontinuous. Since $(E, C, \mathcal{P})$ is locally decomposable, E' is order-convex in E^* and so $F(B) = (B + C') \cap (B - C') = (B + C^*) \cap (B - C^*)$. By Theorem (1.3.1) (b) , $F(B)$ is $\mathcal{P}$-equicontinuous and hence $\beta(E', E)$-bounded.

Conversely, let V be an absolutely convex $\mathcal{P}$-neighbourhood of 0. Then V^{o} is $\mathcal{P}$-equicontinuous subset of E', hence $F(V^{o})$ is $\mathcal{P}$-equicontinuous since E is infrabarrelled and locally o-convex. As $\overline{D(V)} = (F(V^{o}))^{o}$, it follows that $\overline{D(V)}$ is a $\mathcal{P}$-neighbourhood of 0, hence the order-convexity of E' in E^{*} implies that $(E, C, \mathcal{P})$ is locally decomposable by Wong and Ng [1, (3.13), p.38].

(1.3.11) Corollary. <u>Let</u> $(E, C, \mathcal{P})$ <u>be an infrabarrelled, ordered convex space with a countable fundamental system of</u> $\mathcal{P}$-<u>bounded sets in</u> E. <u>If</u> $(E, C, \mathcal{P})$ <u>is locally solid, then so is</u> $(E', C', \beta(E', E))$.

<u>Proof</u>. It is known that an ordered convex space is locally solid if and only if it is both locally o-convex and locally decomposable. According to proposition (1.3.10), $(E', C', \beta(E', E))$ is locally o-convex. Therefore we complete the proof by showing that $(E', C', \beta(E', E))$ is locally decomposable. Since $(E, C, \mathcal{P})$ is locally o-convex, and since $(E', C', \beta(E', E))$ is metrizable and surely bornological, it follows from Proposition (1.3.9) that $(E', C', \beta(E', E))$ is locally decomposable.

(1.3.12) Theorem. <u>Let</u> $(E, C, \mathcal{P})$ <u>be a metrizable ordered convex space such that</u> C <u>is</u> $\mathcal{P}$-<u>complete</u>. <u>If</u> $(E', C', \beta(E', E))$ <u>is locally o-convex then</u> $(E, C, \mathcal{P})$ <u>is locally decomposable, and hence</u> E <u>is</u> $\mathcal{P}$-<u>complete</u>.

<u>Proof</u>. Let $G = C - C$ and let $\mathcal{P}_D$ be the locally decomposable topology on G associated with the relative topology on G induced by $\mathcal{P}$. Then $\mathcal{P}_D$ is metrizable, the embedding mapping J from $(G, \mathcal{P}_D)$ into $(E, \mathcal{P})$ is continuous, and $(G, \mathcal{P}_D)$ is complete by Klee's theorem because C is $\mathcal{P}$-complete. Using a similar argument given in the proof of Proposition (1.3.4), we can show that the embedding map J is nearly open.

According to the open mapping theorem (see Husain [1, Theorem 2, P.36]), J is an open mapping; in other words, for any convex circled $\mathcal{P}$-neighbourhood V of 0 in E, there exists a convex circled $\mathcal{P}$-neighbourhood W of 0 in E such that $W \subseteq D(V)$, it then follows that $G = E$ and $\mathcal{P}_D = \mathcal{P}$. Therefore $(E, C, \mathcal{P})$ is a complete, locally decomposable space, and the proof is complete.

Remark. The preceding theorem is still true when the condition that C be $\mathcal{P}$-complete is replaced by the following:

C *is monotonically sequentially* $\mathcal{P}$-*complete and is* $\mathcal{P}$-*closed*.

Audô-Ellis' theorem says that for an ordered normed space $(E, C, \|\cdot\|)$ in which C is $\|\cdot\|$-complete, C is a strict $\mathcal{B}$-cone in $(E, \|\cdot\|)$ if and only if C' is a normal cone in $(E', \|\cdot\|)$ (i.e., $(E', C', \|\cdot\|)$ is a locally o-convex space). Therefore, Proposition (1.3.10) and Theorem (1.3.12) constitute a generalization of Audô-Ellis' theorem.

(1.3.13) Theorem. *For an ordered convex space* $(E, C, \mathcal{P})$, *if* C *is a generating, then* $\mathcal{P}_{FD} = \mathcal{P}_S = \mathcal{P}_{DF}$.

Proof. We first note that the family $\{V \cap C - V \cap C : V \in \mathcal{U}\}$ is also a neighbourhood base at 0 for $\mathcal{P}_D$. Next we show that

$$\mathcal{P}_F \leq \mathcal{P}_S \leq \mathcal{P}_D . \tag{3.1}$$

The first inequality is obvious since $S(V) \subseteq F(V)$, and the second inequality is true since

$$V_1 \cap C - V_1 \cap C \subseteq S(V)$$

whenever $V_1 + V_1 \subseteq V$.

By inequality (3.1), we immediately obtain

$$\mathcal{P}_{FD} \leq \mathcal{P}_S \leq \mathcal{P}_{DF} .$$

It remains to show that $\mathcal{P}_{DF} \leq \mathcal{P}_{FD}$. Take a $\mathcal{P}_{DF}$- neighbourhood of 0 in E , say $F(V \cap C - V \cap C)$, where $V \in \mathcal{U}$. Then it is not hard to see that

$$F(V) \cap C - F(V) \cap C \subseteq F(V \cap C - V \cap C) .$$

Therefore $F(V \cap C - V \cap C)$ is a $\mathcal{P}_{FD}$-neighbourhood of 0 , and thus $\mathcal{P}_{DF} \leq \mathcal{P}_{FD}$.

(1.3.14) Corollary. Let $(E, C, \mathcal{P})$ be a locally o-convex space for which C is generating, and let Λ be a family of continuous monotone seminorms on E generated $\mathcal{P}$. Then $\mathcal{P}_S = \mathcal{P}_D$, $\mathcal{P}_S$ is determined by the family of seminorms $\{P_D : p \in \Lambda\}$ (or by $\{p_S : p \in \Lambda\}$, $\{p_B : p \in \Lambda\}$) , and the topological dual $(E, \mathcal{P}_S)'$ is the solid subspace of (E^*, C^*) generated by $E' = (E, \mathcal{P})'$. In particular, if $\mathcal{P}$ is locally solid then E' is a solid subspace of (E^*, C^*) .

Proof. This follows from (1.3.13),(1.3.2) and Lemma (1.2.5).

Let $(E, C, \mathcal{P})$ be an ordered convex space such that $E = C - C$ and $E' = C' - C'$. Then $\sigma(E, E')$ is a locally o-convex topology and is determined by the family $\{p_f : f \in C'\}$ of monotone seminorms, where each p_f is defined by

$$p_f(x) = |f(x)| \qquad (x \in E)$$

By (1.3.14), $\sigma_S(E, E')$ is determined by the family $\{p_{f,S} : f \in C'\}$ as well as by $\{p_{f,D} : f \in C'\}$ of seminorms, where $p_{f,S}$ and $p_{f,D}$ are given respectively by

$$p_{f,S}(x) = \inf \{f(u) : u \pm x \in C\}$$

$$p_{f,D}(x) = \inf \{f(u) + f(w) : u,w \in C, x = u - w\} \qquad (x \in E)$$

On the other hand, for any $f \in C'$, let

$$q_f(x) = \sup\{g(x) : g \in E', -f \leq g \leq f\} \qquad (x \in E) .$$

Then the family $\{q_f : f \in C'\}$ of seminorms generates the topology $o(E, E')$ of uniform convergence on all order-intervals in E'. The following result gives a connection between the topologies $o(E, E')$ and $\sigma_S(E, E')$.

(1.3.15) Proposition. Let $(E, C, \mathcal{P})$ be an ordered convex space such that $E = C - C$ and $E' = C' - C'$. For each $f \in C'$, let

$$W_f = \{x \in E : p_f(x) = |f(x)| \leq 1\} \text{ and } V_f = \{x \in E : q_f(x) \leq 1\} .$$

Then the following assertions hold:

(a) $\overline{D(W_f)} = \overline{S(W_f)} = V_f$, hence $q_f \leq p_{f,S}$ and $o(E, E') \leq \sigma_S(E, E')$.

(b) If, in addition, E' is order-convex in (E^*, C^*), then $p_{f,S} = p_{f,D} = q_f$ and thus $\sigma_S(E, E') = o(E, E')$.

Proof. (a) We first note that $D(W_f) \subset S(W_f)$ and that $p_{f,S}$ and $p_{f,D}$ are the gauges of $S(W_f)$ and $D(W_f)$ respectively. For any $g \in E'$ with $-f \leq g \leq f$ and any $u \in C$ with $-u \leq x \leq u$, we have that $|g(x)| \leq f(u)$; it then follows that $q_f(x) \leq p_{f,S}(x)$, and hence that $S(W_f) \subset V_f$. As V_f is the polar of $\{g \in E' : -f \leq g \leq f\} = [-f, f]$, it follows that V_f is $\mathcal{P}$-closed. Therefore $\overline{D(W_f)} \subseteq \overline{S(W_f)} \subseteq V_f$.

In order to verify that $V_f \subseteq \overline{D(W_f)}$, it is sufficient to show, by the bipolar theorem that $(D(W_f))^\circ \subset V_f^\circ = [-f, f]$, where $(D(W_f))^\circ$

is the polar of $D(W_f)$, taken in E' . In fact, if $g \in (D(W_f))^o$, then $|g(x)| \leq p_{f,D}(x)$ for all $x \in E$; in particular,

$$|g(u)| \leq p_{f,D}(u) = f(u) \quad \text{for all} \quad u \in C .$$

which turns out that $g \in [-f, f] = V_f^o$; therefore $(D(W_f))^o \subseteq V_f^o$.
This shows that

$$\overline{D(W_f)} = \overline{S(W_f)} = V_f .$$

(b) If E' is order-convex in (E^*, C^*), then $\sigma_S(E, E')$ is consistent with the dual pair $\langle E, E'\rangle$, and hence $\overline{S(W_f)}$ (resp. $\overline{D(W_f)}$) is the $\sigma_S(E, E')$-closure of $S(W_f)$ (resp. $D(W_f)$) . As $S(W_f)$ and $D(W_f)$ are convex, circled $\sigma_S(E, E')$-neighborhoods of 0 , $p_{f,S}$ and $p_{f,D}$ are the gauges of $S(W_f)$ and $D(W_f)$ resp., we conclude from Schaefer [1 , (1.5) p.40] that $p_{f,S}$ (resp. $p_{f,D}$) is the gauge of $\overline{S(W_f)}$ (resp. $\overline{D(W_f)}$) , and hence from $\overline{D(W_f)} = \overline{S(W_f)} = V_f$ that $q_f = p_{f,S} = p_{f,D}$. Consequently, $\sigma_S(E, E') = o(E, E')$.

(1.3.16) Corollary. <u>Let</u> (E, C) <u>and</u> (G, K) <u>be ordered vector spaces which form a dual pair</u>, <u>let</u> $K = -C^o$ <u>and suppose that</u> $E = C - C$ <u>and</u> $G = K - K$. <u>Then the following assertions hold</u>:

(1) <u>If</u> $\mathcal{P}$ <u>is a locally decomposable topology on</u> E <u>finer than</u> $\sigma(E, G)$, <u>then</u> $\sigma_S(E, G) \leq \mathcal{P}$ <u>and surely</u> $o(E, G) \leq \mathcal{P}$.

(2) <u>If there is an order-interval in</u> G <u>that is not</u> $\sigma(G, E)$-<u>compact</u>, <u>then there does not exist a locally decomposable topology on</u> E <u>which is consistent with</u> $\langle E, G\rangle$.

<u>Proof</u>. (a) Since $\sigma_S(E, G)$ is the smallest locally decomposable

topology on E finer than $\sigma(E, G)$, it follows that $\sigma_S(E, G) \leq \mathcal{P}$, and hence from (1.3.15) (a) that $o(E, G) \leq \mathcal{P}$.

(b) If there exists a locally decomposable topology $\mathcal{P}$ on E consistent with $\langle E, G \rangle$, then $o(E, G) \leq \mathcal{P}$ and thus each order-interval in G is $\sigma(G, E)$-compact.

(1.3.17) Corollary. *Let* (E, C) *and* (G, K) *be ordered vector spaces which form a dual pair. Let* $K = -C^{\circ}$, $E = C - C$ *and* $G = K - K$. *Suppose further that* $\sigma(E, G)$ *is a locally decomposable topology on* E. *Then the following assertions hold:*

(a) *Each order-interval in* (G, K) *is contained in a finite dimensional subspace of* G.

(b) *If* G *contains an order unit, then* E *and* G *are finite dimensional*.

(c) *If there exists a metrizable vector topology* $\mathcal{P}$ *on* G *for which* K *is* $\mathcal{P}$*-complete, then* E *is finite dimensional*.

Proof. We first observe from (1.3.15) that $\sigma(E, G) = o(E, G) = \sigma_S(E, G)$.

(a) For any $f \in K$, there exists $g \in G$ such that for any $h \in G$ with $-f \leq h \leq f$, there is

$$|h(x)| \leq |g(x)| \qquad (x \in E),$$

hence there is a constant α_h such that $h = \alpha_h g$. This means that $[-f, f]$ is contained in a finite dimensional subspace of G.

(b) Follows from the assertion (a).

(c) Let $\{V_n : n \geq 1\}$ be a neighbourhood base at 0 for $\mathcal{P}$ consisting of circled $\mathcal{P}$-closed sets in G such that $V_{n+1} + V_{n+1} \subseteq V_n$ for all $n \geq 1$. Suppose, on the contrary, that E is not finite dimensional. We first choose $f_1 \in V_1 \cap K$, and let G_1 denote the vector subspace of G generated by $[-f_1, f_1]$. By the assertion (a), G_1 is finite dimensional, hence there exists $f_2 \in V_2 \cap K$ such that $f_2 \notin G_1$ because E is not finite dimensional. Let G_2 be the vector subspace of G generated by $[-f_1, f_1] \cup [-f_2, f_2]$. By the assertion (a), G_2 is finite dimensional, so $G_2 \neq G$. After $f_{n-1} \in V_{n-1} \cap K$ $(n \geq 2)$ has been chosen, select $f_{n+1} \in V_{n+1} \cap K$ such that $f_{n+1} \notin G_n$, where G_n is the vector subspace of G generated by $\cup_{k=1}^{n} [-f_k, f_k]$. It is clear that $\{\Sigma_{k=1}^{n} f_k : n \geq 1\}$ is a $\mathcal{P}$-Cauchy sequence in K. The completeness of K implies that there exists $f \in K$ such that

$$f = \Sigma_{k=1}^{\infty} f_k .$$

Clearly $[-f_k, f_k] \subset [-f, f]$ for all $k \geq 1$. Therefore $[-f, f]$ is not contained in a finite dimensional subspace of G which contradicts the assertion (a). Therefore E must be finite dimensional.

Let (E, C) be a Riesz space and let $\mathcal{P}$ be a locally convex topology on E such that the lattice operations are $\mathcal{P}$-continuous. Then it is easily seen that $\mathcal{P}$ is a locally decomposable topology. Therefore (1.3.16) and (1.3.17) are improvements of Peressini [1, (2.8), (2.10), (2.13) and (2.14) of Chap. 3].

In the final chapter we shall study this topology $\sigma_S(E, E')$ in detail, and shall give other characterizations of $\sigma_S(E, E')$ by means of some special type mappings.

CHAPTER II. ORDERS AND TOPOLOGIES ON SPACES CONSISTING OF FAMILIES

In this chapter we study several types of spaces consisting of families, and then consider locally solid or locally o-convex topologies on such spaces. The first important one is the space $\ell^1\langle A, E\rangle$ consisting of elements which are the difference of two positive summable families with index set A. The second type is the space $m_\infty(A, E)$ consisting of elements which are the difference of two positive families with index set A that are majorized by some elements. An interesting locally solid typology $\mathcal{P}_{\varepsilon D}$ on $\ell^1\langle A, E\rangle$ is the locally decomposable topology associated with the relative topology induced by the $\mathcal{P}_\varepsilon$-topology, and then we show that the topological dual of $\ell^1\langle A, E\rangle$ is isomorphic with $m_\infty(A, E')$. Using the natural topology on the bidual E'', we are able to induce a locally solid topology $\mathcal{P}_\infty$ on $m_\infty(A, E)$. If $m_0(A, E)$ denotes the $\mathcal{P}_\infty$-closure of $E^{(A)}$ in $m_\infty(A, E)$, then the topological dual of $m_0(A, E)$ is isomorphic with the space $p_0(A, E')$ consisting of elements which are the difference of two positive families whose partial sums are majorized by some elements.

2.1 Summability of families

In this section we review some basic notions and results on spaces of summability of families, which we shall need in what follows. We assume throughout this section that $(X, \mathcal{P})$ is a locally convex space, that $\mathcal{P}$ is determined by a family P of seminorms, that A is a non-empty index set, and that $\mathcal{F}(A)$ is the directed set consisting of all non-empty finite subsets of A ordered by the set inclusion. Denote by X^A (resp. $X^{(A)}$) the algebraic product (resp. algebraic direct sum) of X with A times, and by $\mathcal{P}^A$ (resp. $\mathcal{P}^{(A)}$) is the product (resp. locally convex direct sum) topology. Elements in X^A will be denoted by (x_i, A) and called <u>families</u> (with index set A) in X, while elements in $X^{(A)}$ will be denoted by $(x_i, (A))$. For each $\alpha \in \mathcal{F}(A)$, we define the map J_α from X into $X^{(A)}$ by setting

$$\pi_i(J_\alpha(x)) = \begin{cases} x & \text{if } i \in \alpha \\ 0 & \text{if } i \notin \alpha \end{cases}$$

where π_i is the i-th projection; in particular, if $\alpha = \{i\}$, then we write J_i for J_α. Clearly, J_α is linear and injective. J_α is referred to as the <u>natural map with respect to</u> α, while J_i is called simply the <u>natural map</u>.

Let $(x_i, A) \in X^A$. Following Pietsch [1], (x_i, A) is said to be <u>weakly summable</u> if for any $f \in X'$, $(\langle x_i, f\rangle, A)$ is a summable family in $\mathbb{R}$, i.e.,

$$\sum_A |\langle x_i, f\rangle| = \sup\{\textstyle\sum_{i\in\alpha} |\langle x_i, f'\rangle| : \alpha \in \mathcal{F}(A)\} < \infty ;$$

(x_i, A) is said to be <u>summable</u> if the net $\{\sum_{i\in\alpha} x_i : \alpha \in \mathcal{F}(A)\}$ in X is $\mathcal{P}$-Cauchy, and (x_i, A) is said to be <u>absolutely summable</u> if for any continuous seminorm p on X, $(p(x_i), A)$ is a summable family in $\mathbb{R}$. (x_i, A) is called a <u>null family</u> in X if for any convex $\mathcal{P}$-neighbourhood V of 0 in X there exists an $\alpha \in \mathcal{F}(A)$ such that

$$x_i \in V \quad \text{for all} \quad i \notin \alpha .$$

It is easily seen that (x_i, A) is weakly summable if and only if the set $\{\sum_{i\in\alpha} x_i : \alpha \in \mathfrak{F}(A)\}$ is $\mathcal{P}$-bounded in view of Pietsch [1 , (1.1.2) p.19] and the fact that $\mathcal{P}$-boundedness and $\sigma(X, X')$-boundedness are the same; also (x_i, A) is a null family in X if and only if for any $p \in P$, the family of numbers $(p(x_i), A)$ is convergent to 0 in the sense of Pietsch [1 , p.21]. By making use of Pietsch [1 , (1.3.6) p.26], summable families in X are weakly summable. The sum of a summable family (x_i, A) in X is defined to be the <u>limit</u> x of the $\mathcal{P}$-Cauchy net $\{\sum_{i\in\alpha} x_i : x \in \mathfrak{F}(A)\}$ which belongs to the completion of X, and we write

$$x = \Sigma_A x_i .$$

It should be noted that our terminology for a summable (resp. absolutely summable) family (x_i, A) differs slightly from that of Schaefer [1 , p.179] in that we require $\{\sum_{i\in\alpha} x_i : \alpha \in \mathfrak{F}(A)\}$ to be convergent in X (resp. $\{\sum_{i\in\alpha} x_i : \alpha \in \mathfrak{F}(A)\}$ to be convergent and for any $p \in P$, $(p(x_i), A)$ to be summable). It $(X, \mathcal{P})$ is complete, then our definition of summability and of absolute summability coincides with that of Schaefer [1 , Prob.23, p.120].

Denote by $\ell^1_w(A, X)$ the vector subspace of X^A consisting of all weakly summable families in X, by $\ell^1(A, X)$ the vector subspace of X^A consisting of all summable families in X, by $\ell^1[A, X]$ the vector subspace of X^A consisting of all absolutely summable families in X, and by $c_0(A, X)$ the vector subspace of X^A consisting of all null families in X. If $A = \mathbb{N}$, then we write $\ell^1_w(X)$ for $\ell^1_w(\mathbb{N}, X)$, $\ell^1(X)$ for $\ell^1(\mathbb{N}, X)$, $\ell^1[X]$ for $\ell^1[\mathbb{N}, X]$, and $c_0(X)$ for $c_0(\mathbb{N}, X)$. It is clear that

$$X^{(A)} \subset \ell^1[A, X] \subset \ell^1(A, X) \subset \ell^1_w(A, X) \quad \text{and} \quad \ell^1(A, X) \subset c_0(A, X) .$$

For any $q \in P$, let $V_q = \{x \in X : q(x) \leqslant 1\}$ and define

(a) $\quad q_{\pi}(x_{\iota}, A) = \sum_A q(x_{\iota})$ $\qquad$ $(x_{\iota}, A) \in \ell^1[A, X];$

(b) $\quad q_w(x_{\iota}, A) = \sup \{ \sum_A |<x_{\iota}, f>| : f \in V_q^{\circ} \}$ $\qquad$ $(x_{\iota}, A) \in \ell_w^1(A, X);$

(b)* $\quad q_{\varepsilon}(x_{\iota}, A) = q_w(x_{\iota}, A)$ $\qquad$ $(x_{\iota}, A) \in \ell^1(A, X);$

(c) $\quad q_{(n)}(x_{\iota}, A) = \sup \{q(x_{\iota}) : \iota \in A\}$ $\qquad$ $(x_{\iota}, A) \in c_0(A, X).$

Clearly q_w, q_{ε}, q_{π} and $q_{(n)}$ are seminorms on their corresponding defined spaces,

$$q_{\varepsilon}(x_{\iota}, A) \leqslant q_{\pi}(x_{\iota}, A) \quad \text{for all} \quad (x_{\iota}, A) \in \ell^1[A, X] \qquad (1.1)$$

Denote by $\mathcal{P}_w$ the locally convex topology on $\ell_w^1(A, X)$ determined by $\{q_w : q \in P\}$, by $\mathcal{P}_{\varepsilon}$ the locally convex topology on $\ell^1(A, X)$ determined by $\{q_{\varepsilon} : q \in P\}$, by $\mathcal{P}_{\pi}$ the locally convex topology on $\ell^1[A, X]$ determined by $\{q_{\pi} : q \in P\}$, and by $\mathcal{P}_{(n)}$ the locally convex topology on $c_0(A, X)$ determined by $\{q_{(n)} : q \in P\}$. Then $\mathcal{P}_{\varepsilon}$ is the relative topology on $\ell^1(A, X)$ induced by $\mathcal{P}_w$, and $\ell^1(A, X)$ is $\mathcal{P}_w$-closed in $\ell_w^1(A, X)$ (see Pietsch [1 , (1.3.3) p.25]. If $(X, \mathcal{P})$ is complete, then so are $(\ell_w^1(A, X), \mathcal{P}_w)$, $(\ell^1(A, X), \mathcal{P}_{\varepsilon})$ and $(\ell^1[A, X], \mathcal{P}_{\pi})$ (see Pietsch [1 , (1.2.4), (1.3.4) and (1.4.3)]. By a similar argument given in the proof of Pietsch [1 , (1.4.3)], $(c_0(A, X), \mathcal{P}_{(n)})$ is complete whenever $(X, \mathcal{P})$ is complete. On the other hand, if $\mathcal{P}$ is metrizable, then so are $\mathcal{P}_w$, $\mathcal{P}_{\varepsilon}$, $\mathcal{P}_{\pi}$ and $\mathcal{P}_{(n)}$. In view of (1.1), the canonical embedding map from $\ell^1[A, X]$ into $\ell^1(A, X)$ is a continuous linear map from $(\ell^1[A, X], \mathcal{P}_{\pi})$ into $(\ell^1(A, X), \mathcal{P}_{\varepsilon})$. The relative topology on $\ell^1(A, X)$ (resp. $c_0(A, X)$) induced by the product topology $\mathcal{P}^A$ is clearly coarser than $\mathcal{P}_{\varepsilon}$ (resp. $\mathcal{P}_{(n)}$) . It is worthwhile to note that if the canonical embedding map from $\ell^1[A, X]$ into $\ell^1(A, X)$ is an algebraic (resp. topological) isomorphism from the first space onto the second space for $A = \mathbb{N}$, then the same is true for any non-empty index set A (see Schaefer [1, p. 182]).

For any $\alpha \in \mathcal{F}(A)$, let k be the number of elements in α. Then we have

$$q_{(n)}(J_\alpha(x)) = q(x) \quad \text{and} \quad q_\varepsilon(J_\alpha(x)) = k\, q(x) = q_\pi(J_\alpha(x)) \qquad (1.2)$$

hence we obtain the following

(2.1.1) Lemma. <u>For any</u> $\alpha \in \mathcal{F}(A)$, <u>the natural map</u> J_α <u>with respect to</u> α <u>is a topological isomorphism from</u> $(X, \mathcal{P})$ <u>into</u> $(c_0(A, X), \mathcal{P}_{(n)})$ <u>as well as into</u> $(\ell^1(A, X), \mathcal{P}_\varepsilon)$ <u>and</u> $(\ell^1[A, X], \mathcal{P}_\pi)$. <u>In particular, if</u> $(X, \|\cdot\|)$ <u>is a normed vector space, then</u> J_ι <u>is an isometry from</u> $(X, \|\cdot\|)$ <u>into the normed space</u> $(c_0(A, X), \mathcal{P}_{(n)})$ <u>as well as into</u> $(\ell^1(A, X), \mathcal{P}_\varepsilon)$ <u>and</u> $(\ell^1[A, X], \mathcal{P}_\pi)$.

The following result deals with the density of $X^{(A)}$ in $(\ell^1[A, X], \mathcal{P}_\pi)$ as well as in $(\ell^1(A, X), \mathcal{P}_\varepsilon)$ and $(c_0(A, X), \mathcal{P}_{(n)})$.

(2.1.2) Lemma. <u>For a locally convex space</u> $(X, \mathcal{P})$, <u>the following assertions hold</u>:

(a) $X^{(A)}$ <u>is dense in</u> $(\ell^1[A, X], \mathcal{P}_\pi)$.

(b) $X^{(A)}$ <u>is dense in</u> $(\ell^1(A, X), \mathcal{P}_\varepsilon)$.

(c) $X^{(A)}$ <u>is dense in</u> $(c_0(A, X), \mathcal{P}_{(n)})$.

<u>Proof</u>. For each $\alpha \in \mathcal{F}(A)$, we define a linear map $T_\alpha : (x_i, A) \to (y_i^{(\alpha)}, A)$ from X^A into $X^{(A)}$ by setting

$$y_i^{(\alpha)} = \begin{cases} x_i & \text{if } i \in \alpha \\ 0 & \text{if } i \notin \alpha . \end{cases}$$

If q is a continuous seminorm on X, then we have

$$(1)\ q_{(n)}((y_i^{(\alpha)}, A) - (x_i, A)) = \sup\{q(x_i) : i \in A\backslash\alpha\} \quad ((x_i, A) \in c_0(A, X))$$

$$(2)\ q_\varepsilon((y_i^{(\alpha)}, A) - (x_i, A)) = \sup\{\textstyle\sum_{A\backslash\alpha} |<x_i, f>| : f \in V_q^o\} \quad ((x_i, A) \in \ell^1(A, X))$$

$$(3)\ q_\pi((y_i^{(\alpha)}, A) - (x_i, A)) = \textstyle\sum_{A\backslash\alpha} q(x_i) \quad ((x_i, A) \in \ell^1[A, X]).$$

Therefore the conclusions (a) and (c) are clear, while the conclusion (b) follows from Pietsch [1 , (1.1.2)] .

A family (x_i', A) in X' is said to be <u>equicontinuous</u> if the set $\{x_i' : i \in A\}$ is an equicontinuous subset of X' .

(2.1.3) Theorem. <u>A linear functional</u> f <u>on</u> $\ell^1[A, X]$ <u>is</u> $\mathcal{P}_\pi$<u>-continuous if and only if there exists a unique equicontinuous family</u> (x_i', A) <u>in</u> X' <u>such that</u>

$$<(x_i, A), f> = \textstyle\sum_A <x_i, x_i'> \quad \underline{\text{for all}} \quad (x_i, A) \in \ell^1[A, X] .$$

<u>Proof</u>. Let S_a denote the vector subspace of X^A consisting of absolutely summable families in the sense of Schaefer [1 , p.150] . It is clear that

$$X^{(A)} \subset S_a \subset \ell^1[A, X] .$$

Since $X^{(A)}$ is dense in $(\ell^1[A, X], \mathcal{P}_\pi)$, it follows that S_a is dense in $(\ell^1[A, X], \mathcal{P}_\pi)$. On the other hand, the topology on S_a in the sense of Schaefer [1 , p.180] is the relative topology on S_a induced by $\mathcal{P}_\pi$. On account of Schaefer [1 , (10.3) p.180], the theorem follows.

A subset M of X' is said to be <u>prenuclear</u> if there exists a $\sigma(X', X)$-closed equicontinuous subset B of X' and a positive Radon measure μ on B such that

$$|\langle x, m\rangle| \leq \int_B |\langle x, x'\rangle| \, d\mu(x') \qquad (x \in X, m \in M) .$$

A family (x_i', A) in X' is said to be <u>prenuclear</u> if the set consisting of the family (x_i', A) is prenuclear.

A similar argument given in the proof of Theorem (2.1.3) yields the following result:

(2.1.4) Theorem. <u>A linear functional f on $\ell^1(A, X)$ is $\mathcal{P}_\varepsilon$-continuous if and only if there exists a unique prenuclear family (x_i', A) in X' such that</u>

$$\langle (x_i', A), f\rangle = \sum_A \langle x_i, x_i'\rangle \qquad \underline{\text{for all}} \quad (x_i, A) \in \ell^1(A, X) .$$

<u>Proof</u>. Let S be the vector subspace of X^A consisting of all summable families in the sense of Schaefer [1 , p.180]. Then

$$X^{(A)} \subset S \subset \ell^1(A, X) .$$

As $X^{(A)}$ is dense in $(\ell^1(A, X), \mathcal{P}_\varepsilon)$, it follows that S is dense in $(\ell^1(A, X), \mathcal{P}_\varepsilon)$. Furthermore, the topology in the sense of Schaefer [1 , p.180] is the relative topology on S induced by $\mathcal{P}_\varepsilon$, the result then follows from Schaefer [1 , (IV. 10.4) p.181] .

(2.1.5) Theorem. <u>A linear functional f on $c_0(A, X)$ is $\mathcal{P}_{(n)}$-continuous if and only if there exists a $\sigma(X', X)$-closed convex circled equicontinuous subset B of X', $\lambda > 0$ and a family (x_i', A) in X' such that</u>

$$\sum_A p_B(x_i') \leq \lambda \quad \underline{\text{and}} \quad \langle (x_i, A), f\rangle = \sum_A \langle x_i, x_i'\rangle \quad \underline{\text{for all}} \ (x_i, A) \in c_0(A, X)$$

where p_B is the norm on the vector subspace $X'(B) = \bigcup_n nB$ induced by B.

Proof. Let q be the gauge of $V = B^o$. The definition of $q_{(n)}$ insures that

$$q(x_j) \leq q_{(n)}(x_i, A) \quad \text{for all} \quad j \in A$$

whenever $(x_i, A) \in c_o(A, X)$. For any $\delta > 0$, $x_j \in (q_{(n)}(x_i, A) + \delta) V$ $(j \in A)$, it follows that

$$|\langle x_j, x_j' \rangle| \leq (q_{(n)}(x_i, A) + \delta) \, p_B(x_j') \quad \text{for all} \quad j \in A ;$$

therefore we obtain the following inequality

$$\sum_A |\langle x_j, x_j' \rangle| \leq (q_{(n)}(x_i, A) + \delta) \sum_A p_B(x_j') \leq \lambda(q_{(n)}(x_i, A) + \delta) \qquad (1.3)$$

As δ was arbitrary, we conclude from (1.3) that (x_i', A) defines a continuous linear functional on $(c_o(A, X), \mathcal{P}_{(n)})$. This proves the condition is sufficient.

To prove the necessity, let q be a continuous seminorm on X such that

$$|f((x_i, A))| \leq q_{(n)}(x_i, A) \quad \text{for all} \quad (x_i, A) \in c_o(A, X) ,$$

let $V_q = \{x \in X : q(x) \leq 1\}$ and let $B = V_q^o$. Then B is a $\sigma(X', X)$-closed convex circled equicontinuous subset of X'. For each $i \in A$, as the natural map J_i is a topological isomorphism from $(X, \mathcal{P})$ into $(c_o(A, X), \mathcal{P}_{(n)})$ it follows that

$$f((x_i, A)) = \sum_A \langle x_i, x_i' \rangle \quad \text{for all} \quad (x_i, A) \in c_o(A, X) .$$

It remains to verify that $\sum_A p_B(x_i') \leq \lambda$ for some $\lambda > 0$.

In fact, for any $\alpha \in \mathcal{F}(A)$ and any $x \in X$, we also have

$$|\sum_{i \in \alpha} \langle x, x_i' \rangle| = |f(J_\alpha x)| \leq q_{(n)}(J_\alpha(x)) = q(x) ,$$

it follows from Pietsch [1 , (1.1.2)] that

$$\sum_{i\in\alpha} |\langle x, x_i'\rangle| \leq 4q(x) .$$

Since $p_B(x_i') = \sup\{|\langle x, x_i'\rangle| : x \in V\}$, we conclude that

$$\sum_{i\in\alpha} p_B(x_i') \leq 4 \quad \text{for all} \quad \alpha \in \mathcal{F}(A) ,$$

and hence that $\sum_A p_B(x_i') \leq 4$.

We have known that $X^{(A)}$ is a vector subspace of $\ell^1[A, X]$ as well as of $c_0(A, X)$, hence we denote by $X_\pi^{(A)}$ (resp. $X_\varepsilon^{(A)}$) the space $X^{(A)}$ equipped with the relative topology induced by $\mathcal{P}_\pi$ (resp. $\mathcal{P}_\varepsilon$) , and by $X_{(n)}^{(A)}$ the space $X^{(A)}$ equipped with the relative topology induced by $\mathcal{P}_{(n)}$. The following result is concerning the continuity of the identity map.

(2.1.6) lemma. <u>Let</u> q <u>be a continuous seminorm on</u> $(X, \mathcal{P})$. <u>Then</u>

$$q_{(n)} \leq q_\varepsilon \leq q_\pi \quad \underline{\text{on}} \quad X^{(A)} .$$

<u>hence the identity maps</u> $X_\pi^{(A)} \to X_\varepsilon^{(A)} \to X_{(n)}^{(A)}$ <u>are all continuous</u>.

<u>Proof</u>. Let V be the unit ball of q in X . For any $(x_i, (A)) \in X^{(A)}$ there exists $\alpha \in \mathcal{F}(A)$ such that $x_j = 0$ for all $j \notin \alpha$. Clearly

$$|\langle x_i, f\rangle| \leq \sum_{i\in\alpha} |\langle x_i, f\rangle| \leq q_\varepsilon(x_i, (A)) \quad \text{for all } f \in V^0 \text{ and } i \in \alpha,$$

it follows that

$$q(x_i) \leq q_\varepsilon(x_i, (A)) \qquad \text{for all} \quad i \in \alpha ,$$

and hence that

$$q_{(n)}(x_i, (A)) \leq q_\varepsilon(x_i, (A)) .$$

The inquality $q_\varepsilon \leq q_\pi$ has been observed, hence the proof is complete.

It is worthwhile to note that the equality

$$q_\varepsilon(x_i, (A)) = \sup\{q(\Sigma_{i\in(A)} c_i x_i) : |c_i| = 1, i \in (A)\}$$

holds for any $(x_i, (A)) \in X^{(A)}$.

2.2 Locally solid topologies on spaces consisting of families

Let (E, C) be an ordered vector space for which C is proper. Denote by C^A the product cone in E^A, and by $C^{(A)} = E^{(A)} \cap C^A$. Then $E^{(A)}$ is always an order-convex subspace of (E^A, C^A). Furthermore, if C is generating, then so is C^A and $E^{(A)} = C^{(A)} - C^{(A)}$, thus $E^{(A)}$ is a solid subspace of (E^A, C^A). It is known from Wong and Ng [1] that if $(E, C, \mathcal{P})$ is locally o-convex (resp. locally decomposable, locally solid), then so are $(E^A, C^A, \mathcal{P}^A)$ and $(E^{(A)}, C^{(A)}, \mathcal{P}^{(A)})$.

For an ordered convex space $(E, C, \mathcal{P})$, we define

(a) $C_\pi(A, E) = C^A \cap \ell^1[A, E]$;

(b) $C_\omega(A, E) = C^A \cap \ell^1_\omega(A, E)$;

(b)* $C_\varepsilon(A, E) = C^A \cap \ell^1(A, E)$;

(c) $C_{(n)}(A, E) = C^A \cap c_o(A, E)$.

Then $C_\pi(A, E)$, $C_\varepsilon(A, E)$ and $C_\omega(A, E)$ are cones in $\ell^1_\omega(A, E)$, and

$C_\pi(A, E) = C_\varepsilon(A, E) \cap \ell^1[A, E] = C_\omega(A, E) \cap \ell^1[A, E] = C_{(n)}(A, E) \cap \ell^1[A, E]$;

$C_\varepsilon(A, E) = C_\omega(A, E) \cap \ell^1(A, E) = C_{(n)}(A, E) \cap \ell^1(A, E)$.

If q is a continuous monotone seminorm on E, then q_π and $q_{(n)}$ are monotone on $\ell^1[A, E]$ and $c_0(A, E)$ respectively, and the polar V_q^o, taken in E', of the unit ball $V_q = \{x \in E, q(x) < 1\}$ is positively dominated in (E', C'). It follows that

$$q_\omega(u_i, A) = \sup\{\sum_A \langle u_i, u'\rangle : u' \in V_q^o \cap C'\} \quad \text{for all } (u_i, A) \in C_\omega(A, E),$$

and hence q_ω is monotone. Furthermore, we have

(2.2.1) Lemma. <u>If</u> $(E, C, \mathcal{P})$ <u>is a locally</u> o-<u>convex space</u>, <u>then so are</u> $(\ell^1[A, E], C_\pi(A, E), \mathcal{P}_\pi)$, $(\ell^1(A, E), C_\varepsilon(A, E), \mathcal{P}_\varepsilon)$, $(\ell^1_\omega(A, E), C_\omega(A,E), \mathcal{P}_\omega)$ <u>and</u> $(c_0(A, E), C_{(n)}(A, E), \mathcal{P}_{(n)})$. <u>Moreover, if</u> q <u>is a continuous monotone seminorm on</u> E, <u>then we have, for any</u> $(u_i, A) \in C_\omega(A, E)$, <u>that</u>

$$\begin{aligned} q_\omega(u_i, A) &= \sup\{\sum_A \langle u_i, u'\rangle : u' \in V_q^o \cap C'\} \\ &= \sup\{q(\sum_{i\in\alpha} u_i) : \alpha \in \mathcal{F}(A)\}, \end{aligned} \tag{2.1}$$

<u>where</u> V_q^o <u>is the polar</u>, <u>taken in</u> E', <u>of</u> $V_q = \{x \in E : q(x) < 1\}$.

<u>Proof</u>. In view of the preceding remark, it is sufficient to verify the last equality of (2.1). It is clear that

$$q_\omega(u_i, A) \leq \sup\{q(\sum_{i\in\alpha} u_i) : \alpha \in \mathcal{F}(A)\}. \tag{2.2}$$

On the other hand, for any $\alpha \in \mathcal{F}(A)$, the $\sigma(E', E)$-compactness of $V_q^o \cap C'$ insures that there exists $h_\alpha \in V_q^o \cap C'$ such that

$$q(\sum_{i\in\alpha} u_i) = \sum_{i\in\alpha} \langle u_i, h_\alpha\rangle. \tag{2.3}$$

Since $u_i \in C$ and since $h_\alpha \in V_q^\circ \cap C'$, it follows that

$$\sum_{i\in\alpha} \langle u_i , h_\alpha \rangle \leqslant \sum_A \langle u_i , h_\alpha \rangle \leqslant q_\omega(u_i , A) ,$$

and hence from (2.3) that

$$\sup\{q(\sum_{i\in\alpha} h_i) : \alpha \in \mathcal{F}(A)\} \leqslant q_\omega(u_i , A) .$$

Combining this with (2.2), we get the required equality.

In particular, we have, for any $(u_i , (A)) \in C^{(A)}$, that

$$q_\omega(u_i , (A)) = q(\sum_{i\in(A)} u_i) .$$

Let $(E, C, \mathcal{P})$ be a locally o-convex space, let $\mathcal{P}$ be determined by a family P of monotone seminorms, and suppose that

$$\ell^1_a\langle A, E\rangle = C_\pi(A, E) - C_\pi(A, E) .$$

Then $\ell^1_a\langle A, E\rangle$ is a solid subspace of (E^A , C^A) as well as of $(\ell^1[A,E], C_\pi(A,E))$. In view of Lemma (2.2.1) and a well-known result, $(\ell^1_a\langle A, E\rangle, C_\pi(A, E))$ equipped with the relative topology induced by $\mathcal{P}_\pi$, is a locally o-convex space. Denoting by $\mathcal{P}_{\pi D}$ the locally decomposable topology on $\ell^1_a\langle A, E\rangle$ associated with the relative topology induced by $\mathcal{P}_\pi$; then $(\ell^1_a\langle A, E\rangle, C_\pi(A, E), \mathcal{P}_{\pi D})$ is a locally solid space, and $\mathcal{P}_{\pi D}$ is determined by the family $\{q_{\pi D} : q \in P\}$ of seminorms, where each $q_{\pi D}$ is given by

$$\text{(a)}\quad q_{\pi D}(x_i , A) = \inf\{q_\pi(u_i , A) + q_\pi(v_i , A) : (u_i , A), (v_i , A) \in C_\pi(A, E), \\ (u_i , A) - (v_i , A) = (x_i , A)\} .$$

Also $\mathcal{P}_{\pi D}$ is determined by $\{q_{\pi S} : q \in P\}$ as well as $\{q_{\pi B} : q \in P\}$, where $q_{\pi S}$ and $q_{\pi B}$ are given by

(b) $q_{\pi S}(x_i, A) = \inf\{q_\pi(u_i, A) : (u_i, A) \in C_\pi(A, E),$
$(u_i, A) \pm (x_i, A) \in C_\pi(A, E)\}$,

(c) $q_{\pi B}(x_i, A) = \inf\{\max(q_\pi(w_i, A), q_\pi(v_i, A)) : (w_i, A), (v_i, A) \in C_\pi(A,E),$
$-w_i \leq x_i \leq u_i (\forall i \in A)\}$.

If $(E, C, \mathcal{P})$ is a <u>metrizable</u> locally solid space, then $\ell^1_a\langle\mathbb{N}, E\rangle = \ell[\mathbb{N}, E]$ and $\mathcal{P}_{\pi D} = \mathcal{P}_\pi$ as shown by the following result whose proof is routine and therefore will be omitted.

(2.2.2) Lemma. <u>If</u> $(E, C, \mathcal{P})$ <u>is a metrizable locally solid space, then so is</u> $(\ell^1[\mathbb{N}, E], C_\pi(\mathbb{N}, E), \mathcal{P}_\pi)$, <u>consequently</u> $\ell^1_a\langle\mathbb{N}, E\rangle = \ell^1[\mathbb{N}, E]$ <u>and</u> $\mathcal{P}_{\pi D} = \mathcal{P}_\pi$.

Let $(E, C, \mathcal{P})$ be a locally o-convex space, let $\mathcal{P}$ be determined by a family P of monotone seminorms, and suppose that

$$\ell^1_\omega\langle A, E\rangle = C_\omega(A, E) - C_\omega(A, E) .$$

Then $\ell^1_\omega\langle A, E\rangle$ is a solid subspace of (E^A, C^A) as well as of $(\ell^1_\omega\langle A, E\rangle, C_\omega(A,E))$. In view of Lemma (2.2.1) and a well-known result, $(\ell^1_\omega\langle A, E\rangle, C_\omega(A, E))$ equipped with the relative topology induced by $\mathcal{P}_\omega$, is a locally o-convex space. If we denote by $\mathcal{P}_{\omega D}$ the locally decomposable topology on $\ell^1_\omega\langle A, E\rangle$ associated with the relative topology induced by $\mathcal{P}_\omega$, then $(\ell^1_\omega\langle A, E\rangle$, $C_\omega(A, E), \mathcal{P}_{\omega D})$ is a locally solid space by (1.3.14) , and $\mathcal{P}_{\omega D}$ is determined by the family $\{q_{\omega D} : q \in P\}$ of seminorms, where each $q_{\omega D}$ is given by

(a) $q_{\omega D}(x_i, A) = \inf\{q_\omega(u_i, A) + q_\omega(v_i, A) : (u_i, A), (v_i, A) \in C_\omega(A, E),$
$(u_i, A) - (v_i, A) = (x_i, A)\}$.

Also $\mathscr{P}_{\omega D}$ is determined by $\{q_{\omega S} : q \in P\}$ as well as by $\{q_{\omega B} : q \in P\}$, where $q_{\omega S}$ and $q_{\omega B}$ are given by

$$\text{(b)}\quad q_{\omega S}(x_i , A) = \inf\{q_\omega(u_i , A) : (u_i , A) \in C_\omega(A, E), (u_i , A) \pm (x_i , A) \in C_\omega(A, E)\} ,$$

$$\text{(c)}\quad q_{\omega B}(x_i , A) = \inf\{\max(q_\omega(y_i , A), q_\omega(z_i , A)) : (y_i , A), (z_i , A) \in C_\omega(A,E), -y_i \leq x_i \leq z_i \ (\forall_i \in A)\} .$$

If $(E, C, \mathscr{P})$ is a locally convex Riesz space, then so is $(\ell^1_\omega\langle A, E\rangle, C_\omega(A, E), \mathscr{P}_{\omega D})$.

Denote by $\ell^1\langle A, E\rangle$ the subspace of $\ell^1(A, E)$ generated by $C_\varepsilon(A, E)$, that is

$$\ell^1\langle A, E\rangle = C_\varepsilon(A, E) - C_\varepsilon(A, E) .$$

Then $\ell^1\langle A, E\rangle \subset \ell^1_\omega\langle A, E\rangle$, $\ell^1\langle A, E\rangle$ is a solid subspace of (E^A, C^A) as well as of $(\ell^1(A, E), C_\varepsilon(A, E))$ and $(\ell^1_\omega\langle A, E\rangle, C_\omega(A, E))$. Since $\mathscr{P}_\varepsilon$ is the relative topology of $\mathscr{P}_\omega$, it follows from (2.1.1) that $(\ell^1\langle A, E\rangle, C_\varepsilon(A, E))$ equipped with the relative topology induced by $\mathscr{P}_\varepsilon$ is a locally o-convex space, and hence that the locally decomposable topology on $\ell^1\langle A, E\rangle$ associated with the relative topology induced by $\mathscr{P}_\varepsilon$, denoted by $\mathscr{P}_{\varepsilon D}$, is the relative topology induced by $\mathscr{P}_{\omega D}$. Therefore $\mathscr{P}_{\varepsilon D}$ is determined by $\{q_{\varepsilon D} : q \in P\}$ as well as $\{q_{\varepsilon S} : q \in P\}$ and $\{q_{\varepsilon B} : q \in P\}$, where

$$q_{\varepsilon D} = q_{\omega D} , \quad q_{\varepsilon S} = q_{\omega S} \quad \text{and} \quad q_{\varepsilon B} = q_{\omega B} \quad \text{on} \quad \ell^1\langle A, E\rangle .$$

As $q_\varepsilon \leq q_\pi$ on $\ell^1[A, E]$, $q_{\varepsilon D} = q_\varepsilon$ on $C_\varepsilon(A, E)$ and $q_{\pi D} = q_\pi$ on $C_\pi(A, E)$, it follows that

$$q_{\varepsilon D} \leq q_{\pi D} , \quad q_{\varepsilon S} \leq q_{\pi S} \quad \text{and} \quad q_{\varepsilon B} \leq q_{\pi B} \quad \text{on} \quad \ell^1_a\langle A, E\rangle .$$

If $(E, C, \mathcal{P})$ is a locally convex Riesz space, then so is $(\ell^1\langle A, E\rangle, C_\varepsilon(A, E), \mathcal{P}_{\varepsilon D})$. Since $\mathcal{P}_{\omega D}$ is finer than the relative topology on $\ell^1_\omega\langle A, E\rangle$ induced by $\mathcal{P}_\omega$ and since $\ell^1(A, E)$ is $\mathcal{P}_\omega$-closed in $\ell^1_\omega(A, E)$, it follows that $\ell^1\langle A, E\rangle$ is $\mathcal{P}_{\omega D}$-closed in $\ell^1_\omega\langle A, E\rangle$. This remark makes the following result clear.

(2.2.3) Lemma. *For a locally o-convex space* $(E, C, \mathcal{P})$, $\mathcal{P}_{\varepsilon D}$ *coincides with the relative topology on* $\ell^1\langle A, E\rangle$ *induced by* $\mathcal{P}_{\omega D}$, *and* $\ell^1\langle A, E\rangle$ *is a* $\mathcal{P}_{\omega D}$*-closed solid subspace of* $(\ell^1_\omega\langle A, E\rangle, C_\omega(A, E))$.

If C is $\mathcal{P}$-closed, then $C_\omega(A, E)$ is $\mathcal{P}_\omega$-closed and $C_\varepsilon(A, E)$ is $\mathcal{P}_\varepsilon$-closed. Moreover, we have

(2.2.4) Lemma. *Let* $(E, C, \mathcal{P})$ *be a Fréchet locally solid space. If* C *is* $\mathcal{P}$*-closed, then* $(\ell^1_\omega\langle A, E\rangle, C_\omega(A, E), \mathcal{P}_{\omega D})$ *and* $(\ell^1\langle A, E\rangle, C_\varepsilon(A, E), \mathcal{P}_{\varepsilon D})$ *are complete.*

Proof. In view of (2.2.3), it is sufficient to verify the completeness of $(\ell^1_\omega\langle A, E\rangle, C_\omega(A, E), \mathcal{P}_{\omega D})$. In view of Pietsch [1, (1.2.4)], $(\ell^1_\omega(A, E), \mathcal{P}_\omega)$ is a Fréchet space, hence $C_\omega(A, E)$ is $\mathcal{P}_\omega$-complete. Since the relatives on $C_\omega(A, E)$ induced by $\mathcal{P}_\omega$ and by $\mathcal{P}_{\omega D}$ coincide, it follows that $C_\omega(A, E)$ is monotonically sequentially complete for $\mathcal{P}_{\omega D}$, and hence from a result of Jameson [1] that $(\ell^1_\omega\langle A, E\rangle, C_\omega(A, E), \mathcal{P}_{\omega D})$ is complete.

(2.2.5) Corollary. *Let* $(E, C, \mathcal{P})$ *be a Fréchet locally solid space and* C $\mathcal{P}$*-closed. If* $C_\omega(A, E)$ *is generating, then* $(\ell^1_\omega(A, E), C_\omega(A, E), \mathcal{P}_\omega)$ *is a Fréchet locally solid space; if* $C_\varepsilon(A, E)$ *is generating, then* $(\ell^1(A, E), C_\varepsilon(A, E), \mathcal{P}_\varepsilon)$ *is a locally solid space.*

Proof. The identity map I_A from $(\ell^1_\omega\langle A, E\rangle, \mathcal{P}_{\omega D})$ into $(\ell^1_\omega(A, E), \mathcal{P}_\omega)$ is a continuous surjective linear map. As $(\ell^1_\omega\langle A, E\rangle, \mathcal{P}_{\omega D})$ and $(\ell^1_\omega(A, E), \mathcal{P}_\omega)$ are complete, it follows that I_A is open, and hence that $\mathcal{P}_\omega = \mathcal{P}_{\omega D}$. The proof of the second assertion is similar.

If C is generating, then $E^{(A)} \subset \ell^1\langle A, E\rangle$. Furthermore, $E^{(A)}$ is dense in $(\ell^1\langle A, E\rangle, \mathcal{P}_{\varepsilon D})$ as the following result shows.

(2.2.6) Lemma. Let $(E, C, \mathcal{P})$ be a locally o-convex space. If C is generating, then $E^{(A)}$ is dense in $(\ell^1\langle A, E\rangle, C_\varepsilon(A, E), \mathcal{P}_{\varepsilon D})$, and

$$q_{\varepsilon D}(J_\alpha(x)) \leq k\, q_D(x) \; ; \; q_{\varepsilon S}(J_\alpha(x)) \leq k\, q_S(x) \; ; \; q_{\varepsilon B}(J_\alpha(x)) \leq k\, q_B(x) ,$$

where $\alpha \in \mathfrak{F}(A)$, k is the number of elements in α, J_α is the natural map w.r.t. α, and q is a continuous monotone seminorm on E.

Proof. Let q be a continuous monotone seminorm on E and $(x_i, A) \in \ell^1\langle A, E\rangle$. For any $\delta > 0$ there exist $(u_i, A), (w_i, A) \in C_\varepsilon(A, E)$ with $(x_i, A) = (u_i, A) - (w_i, A)$ such that

$$q_\varepsilon(u_i, A) + q_\varepsilon(w_i, A) < q_{\varepsilon D}(x_i, A) + \delta .$$

As (u_i, A) and (w_i, A) are positive summable families in E, there is $\gamma \in \mathfrak{F}(A)$ such that

$$q\Big(\sum_{i\in\beta} u_i\Big) < \delta/2 \quad \text{and} \quad q\Big(\sum_{i\in\beta} w_i\Big) < \delta/2$$

whenever $\beta \in \mathfrak{F}(A)$ with $\beta \cap \gamma = \phi$, hence

$$\sup\{q\Big(\sum_{i\in\alpha} u_i\Big) : \alpha \in \mathfrak{F}(A\backslash\gamma)\} \leq \delta/2 \qquad (2.4)$$
$$\sup\{q\Big(\sum_{i\in\alpha} w_i\Big) : \alpha \in \mathfrak{F}(A\backslash\gamma)\} \leq \delta/2$$

For each $\beta \in \mathfrak{F}(A)$, as in the proof of (2.1.2), the map $T_\beta : (x_i, A) \to (y_i^{(\beta)}, A)$, defined by

$$y_i^{(\beta)} = \begin{cases} x_i & \text{if } \iota \in \beta \\ 0 & \text{if } \iota \notin \beta , \end{cases}$$

is a linear map from $\ell^1\langle A, E\rangle$ into $E^{(A)}$. Therefore the formula (2.4) shows that

$$q_\varepsilon(u_i, A) - T_\gamma(u_i, A) \leq \delta/2 \quad \text{and} \quad q_\varepsilon((w_i, A) - T_\gamma(w_i, A)) \leq \delta/2. \quad (2.5)$$

On the other hand, $(u_i - u_i^{(\gamma)}, A) = (u_i, A) - T_\gamma(u_i, A)$ and $(w_i - w_i^{(\gamma)}, A) = (w_i, A) - T_\alpha(w_i, A)$, in $C_\varepsilon(A, E)$, are such that

$$(x_i, A) - T_\gamma(x_i, A) = (u_i - u_i^{(\gamma)}, A) - (w_i - w_i^{(\gamma)}, A) ,$$

it follows from the definition of $q_{\varepsilon,D}$ and (2.5) that

$$q_{\varepsilon D}((x_i, A) - T_\gamma(x_i, A)) \leq q_\varepsilon(u_i - u_i^{(\gamma)}, A) + q_\varepsilon(w_i - w_i^{(\gamma)}, A) \leq \delta .$$

Therefore $E^{(A)}$ is dense in $(\ell^1\langle A, E\rangle, C_\varepsilon(A, E), \mathcal{P}_{\varepsilon D})$.

Finally, by the formula (1.2) in the previous section, we have

$$\begin{aligned} q_D(x) &= \inf\{q(u) + q(w) : u, w \in C , \ u - w = x\} \\ &= k^{-1} \inf\{q_\varepsilon(J_\alpha(u)) + q_\varepsilon(J_\alpha(w)) : J_\alpha(u) - J_\alpha(w) = J_\alpha(x)\} \\ &\geq k^{-1} q_{\varepsilon D}(J_\alpha(x)) . \end{aligned}$$

By a similar argument, we can obtain the other two inequalities.

In view of (2.2.3) and (2.2.6), $E^{(A)}$ is, in general, not dense in $(\ell^1_\omega\langle A, E\rangle, C_\omega(A, E), \mathcal{P}_{\omega D})$. On the other hand, since $\ell^1\langle A, E\rangle$ is $\mathcal{P}_{\omega D}$-closed in $\ell^1_\omega\langle A, E\rangle$, $\mathcal{P}_{\varepsilon D}$ is the relative topology induced by $\mathcal{P}_{\omega D}$, and since $E^{(A)}$ is $\mathcal{P}_{\varepsilon D}$-dense in $\ell^1\langle A, E\rangle$, it follows that $\ell^1\langle A, E\rangle$ is the $\mathcal{P}_{\omega D}$-closure of $E^{(A)}$ in $\ell^1_\omega\langle A, E\rangle$. Therefore we obtain:

(2.2.7) Lemma. Let $(E, C, \mathcal{P})$ be a locally o-convex space. If C is generating, then $\ell^1\langle A, E\rangle$ is the $\mathcal{P}_{\omega D}$-closure of $E^{(A)}$ in $\ell^1_\omega\langle A, E\rangle$.

The remainder of this section is devoted to a study of other typed spaces consisting of families which has, in some sense, the majorized property. Let us define

$$m^+_\infty(A, E) = \{(u_i, A) \in C^A : \exists\, u \in C \text{ such that } u_i \leq u \;\forall\, i \in A\}$$
$$m_\infty(A, E) = m^+_\infty(A, E) - m^+_\infty(A, E) .$$

Then $m^+_\infty(A, E)$ is a subcone of C^A , hence $m_\infty(A, E)$ is a solid subspace of (E^A, C^A) . If C is generating, then $E^{(A)} \subset m_\infty(A, E)$.

If q is a continuous monotone seminorm on E , then the functional λ_q defined by

$$\lambda_q(u_i , A) = \inf\{q(u) : u \in C,\ u_i \leq u \;\forall\, i \in A\} \quad \text{for all } (u_i ,A) \in m^+_\infty(A, E)$$

is a monotone sublinear functional on $m^+_\infty(A, E)$, and has the following property:

$$\lambda_q(J_\alpha(u)) = q(u) \quad \text{for all } u \in C \text{ and } \alpha \in \mathcal{F}(A) .$$

According to (1.2.6), λ_q induces the following monotone seminorm on $m_\infty(A, E)$

(a) $$q_\infty(x_i , A) = p_{\lambda_q}(x_i , A)$$
$$= \inf\{\lambda_q(u_i , A) + \lambda_q(w_i , A) : (u_i , A), (w_i , A) \in m^+_\infty(A, E),$$
$$(u_i , A) - (w_i , A) = (x_i , A)\}$$

such that

$$q_\infty = q_{\infty D} \quad \text{and} \quad q_\infty(u_i , A) = \lambda_q(u_i , A) \quad \text{for all } (u_i , A) \in m^+_\infty(A, E)$$

Denoting

$$V^+_{\infty D} = \{(u_i , A) \in m^+_\infty(A, E) : \lambda_q(u_i , A) < 1\} ;$$
$$V_\infty = \{(x_i , A) \in m_\infty(A, E) : q_\infty(x_i , A) < 1\} .$$

Lemma (1.2.6) tells us that

$$V^+_{\infty D} = V_\infty \cap m^+_\infty(A, E) \quad \text{and} \quad V_\infty = D(V_\infty) = co((-V^+_{\infty D}) \cup V^+_{\infty D}) .$$

Also, the gauges of $S(V_\infty)$ and $B(V_\infty)$ are given respectively by

(b) $q_{\infty S}(x_i , A) = \inf\{\lambda_q(u_i , A) : (u_i , A) \in m^+_\infty(A, E),$
$-(u_i , A) \leq (x_i , A) \leq (u_i , A)\}$

(c) $q_{\infty B}(x_i , A) = \inf\{\max(\lambda_q(u_i , A), \lambda_q(w_i , A)) : (u_i , A), (w_i , A) \in m^+_\infty(A, E),$
$-(u_i , A) \leq (x_i , A) \leq (w_i , A)\} .$

$q_{\infty S}$ is a Riesz seminorm on $m_\infty(A, E)$ and

$q_{\infty S}(u_i , A) = q_{\infty B}(u_i , A) = q_\infty(u_i , A) = \lambda_q(u_i , A)$ for all $(u_i , A) \in m^+_\infty(A, E)$.

Clearly, a family (x_i , A) in E belongs to $m_\infty(A, E)$ if and only if there exists an $u \in C$ such that

$$-u \leq x_i \leq u \quad \text{for all} \quad i \in A .$$

If we define the <u>diagonal map</u> J_∞ from E into E^A , i.e.,

$$J_\infty(x) = (x_i , A) \quad \text{if} \quad x_i = x \quad \text{for all } i \in A \quad (x \in E) ,$$

then J_∞ is positive, linear and injective, $J_\alpha \leq J_\infty$ for all $\alpha \in \mathfrak{F}(A)$, and $J_\infty(C) \subset m^+_\infty(A, E)$, thus $J_\infty(E) \subset m_\infty(A, E)$ whenever C is generating. Furthermore, we have:

(2.2.8) Lemma. <u>Let</u> $(E, C, \mathcal{P})$ <u>be an ordered convex space and</u> q <u>a monotone continuous seminorm on</u> E . <u>Then the following statements hold</u>:

(1) $\lambda_q(J_\infty(u)) = q(u) = \lambda_q(J_\alpha(u))$ <u>for all</u> $u \in C$ <u>and</u> $\alpha \in \mathfrak{F}(A)$

(2) $q_{\infty S}(x_i , A) = \inf\{q(u) : u \in C , -u \leq x_i \leq u \ \forall i \in A\}$ <u>and</u> (2.6)

$q_{\infty B}(x_i , A) = \inf\{\max(q(u), q(w)) : u, w \in C, -u \leq x_i \leq w \ \forall i \in A\}$. (2.7)

(3) <u>If</u> C <u>is generating, then we have, for any</u> $x \in E$ <u>and</u> $\alpha \in \mathfrak{F}(A)$ <u>, that</u>

(i) $q_\infty(J_\alpha(x)) \leq q_D(x)$ <u>and</u> $q_\infty(J_\infty(x)) \leq q_D(x)$,

(ii) $q_{\infty S}(J_\alpha(x)) = q_{\infty S}(J_\infty(x)) = q_S(x)$,

(iii) $q_{\infty B}(J_\alpha(x)) = q_{\infty B}(J_\infty(x)) = q_B(x)$.

(4) <u>If</u> (E, C) <u>is an order complete Riesz space, then</u>

$$q_{\infty S}(x_\iota , A) = q(\sup_A |x_j|) \quad \underline{\text{for all}} \quad (x_\iota , A) \in m_\infty(A, E) .$$

<u>Proof</u>. (1) This follows from the definition of J_∞ and of λ_q .

(2) The proof of (2.7) is similar to that of (2.6), hence we only verify the equality (2.6). For any $u \in C$ with $-u \leq x_\iota \leq u$ for all $\iota \in A$, we have

$$-J_\infty(u) \leq (x_\iota , A) \leq J_\infty(u) ,$$

hence

$$q_{\infty S}(x_\iota , A) \leq \lambda_q(J_\infty(u)) = q(u) ;$$

as $u \in C$ was arbitrary, we have

$$q_{\infty S}(x_\iota , A) \leq \inf\{q(u) : u \in C , -u \leq x_\iota \leq u \ \forall \iota \in A\} .$$

On the other hand, for any $\delta > 0$ there exists $(u_\iota , A) \in m_\infty^+(A, E)$ such that

$$-(u_\iota , A) \leq (x_\iota , A) \leq (u_\iota , A) \text{ and } \lambda_q(u_\iota , A) < q_{\infty S}(x_\iota , A) + \delta/2. \qquad (2.8)$$

By the definition of λ_q , there exists an $u \in C$ such that

$$u_\iota \leq u \ (\forall \iota \in A) \text{ and } q(u) < \lambda_q(u_\iota , A) + \delta/2 . \qquad (2.9)$$

From (2.8) and (2.9), we conclude that there exists an $u \in C$ with $-u \leq x_\iota \leq u$ $(\iota \in A)$ such that

$$\inf\{q(w) : w \in C, -w \leq x_\iota \leq w \ \forall \iota \in A\} \leq q(u) < q_{\infty S}(x_\iota , A) + \delta .$$

As δ was arbitrary, it follows that

$$\inf\{q(w) : w \in C , -w \leq x_i \leq w \ \forall i \in A\} \leq q_{\infty S}(x_i , A) .$$

Therefore we get the equality (2.6).

(3) The equalities (ii) and (iii) follow from (2.6) and (2.7) respectively, while (i) is clear by making use of the definition of q_∞ and of J_α, J_∞ .

(4) If E is an order complete Riesz space, then for any $(x_i , A) \in m_\infty(A, E)$, $\sup|x_j|$ exists in E and $-\sup|x_j| \leq x_i \leq \sup|x_j|$ $(\forall i \in A)$, therefore

$$q_{\infty S}(x_i , A) \leq q(\sup|x_j|).$$

On the other hand, for any $u \in C$ with $-u \leq x_i \leq u$ $(i \in A)$, we have $\sup|x_j| \leq u$, and thus $q(\sup|x_j|) \leq q(u)$; as $u \in C$ was arbitrary, it follows that

$$q(\sup|x_j|) \leq q_{\infty S}(x_i , A)$$

and hence we get the required equality.

Let $(E, C, \mathscr{P})$ be a locally o-convex space and let $\mathscr{P}$ be determined by a family P of monotone seminorms. Denote by $\mathscr{P}_\infty$ the topology on $m_\infty(A, E)$ determined by $\{q_\infty : q \in P\}$. Then $(m_\infty(A, E), m_\infty^+(A, E), \mathscr{P}_\infty)$ is a locally solid space and $\mathscr{P}_\infty$ is also determined by $\{q_{\infty S} : q \in P\}$ as well as by $\{q_{\infty B} : q \in P\}$. If $(E, C, \mathscr{P})$ is metrizable or normable, then so is $(m_\infty(A, E), m_\infty^+(A, E), \mathscr{P}_\infty)$. If C is $\mathscr{P}$-closed, then $m_\infty^+(A, E)$ is $\mathscr{P}_\infty$-closed. If $(E, C, \mathscr{P})$ is a locally convex Riesz space, then so is $(m_\infty(A, E), m_\infty^+(A, E), \mathscr{P}_\infty)$. If $(E, C, \mathscr{P})$ is a Fréchet locally solid space for which C is closed, then

$(m_\infty(A, E), m_\infty^+(A, E), \mathcal{P}_\infty)$ is a Fréchet locally solid space. (see Walsh [1, (2.3.2)] .

Let C be generating. Denote by $m_0(A, E)$ the $\mathcal{P}_\infty$-closure of $E^{(A)}$ in $m_\infty(A, E)$, by $\mathcal{P}_{\infty 0}$ the relative topology on $m_0(A, E)$ induced by $\mathcal{P}_\infty$, and by $m_0^+(A, E)$ the relative cone induced by $m_\infty^+(A, E)$, i.e.

$$m_0^+(A, E) = m_\infty^+(A, E) \cap m_0(A, E) .$$

Then $(m_0(A, E), m_0^+(A, E), \mathcal{P}_{\infty 0})$ is a locally o-convex space, which is called the <u>space of null-majorized families</u>. It should be noted that $m_0^+(A, E)$ may not be generating. If $(E, C, \mathcal{P})$ is a locally convex Riesz space, then $m_0(A, E)$ is a solid subspace of $(m_\infty(A, E), m_\infty^+(A, E))$ because $E^{(A)}$ is solid in $m_\infty(A, E)$ and the closure of a solid set in a locally convex Riesz space is solid, hence $(m_0(A, E), m_0^+(A, E), \mathcal{P}_{\infty 0})$ is a locally convex Riesz space.

In order to induce another typed space consisting of families, we require the following terminology and some well-known facts. Let $(E, C, \mathcal{P})$ be an ordered convex space, let $\mathcal{P}$ be determined by a family S of seminorms on E , let E'' be the bidual of E , and let C'' be the bidual cone of C , i.e., C'' is the set consisting of all positive $\beta(E', E)$-continuous linear functionals on E' . Then E is regarded as a vector subspace of E'' and C is identified with a subset of C'' , hence

$$y'' \leq x \leq z'' \quad \text{in} \quad E'' \quad \text{is meant} \quad z'' - x \in C'' \quad \text{and} \quad x - y'' \in C''$$

where $x \in E$ and $y'', z'' \in E''$. For any $q \in S$, let V be the unit ball of q in E , let V^{oo} be the bipolar of V taken in E'' , and let q'' be the gauge of V^{oo} . Then

$$q''(x'') = \sup\{\langle f, x''\rangle : f \in V^o\} \qquad (x'' \in E'') ,$$

and hence

$$q''(x) = q(x) \qquad \text{for all} \quad x \in E .$$

$\{q'' : q \in S\}$ defines the natural topology $\mathcal{P}''$ on E''. If q is monotone, then it is easily seen that

$$q''(u'') = \sup\{\langle f, u''\rangle : f \in V^o \cap C'\} \quad \text{for all} \quad u'' \in C'' ,$$

hence q'' is monotone. From this we conclude that if $(E, C, \mathcal{P})$ is a locally o-convex space, then so is $(E'', C'', \mathcal{P}'')$.

Let us now define

$$m^+_{\infty,2}(A, E) = \{(u_\iota, A) \in C^A : \exists\, u'' \in C'' \text{ such that } o \leq u_\iota \leq u'' \text{ in } E''\ \forall \iota \in A\},$$
$$m_{\infty,2}(A, E) = m^+_{\infty,2}(A, E) - m^+_{\infty,2}(A, E) .$$

Then $m^+_{\infty,2}(A, E)$ is a subcone of C^A, hence $m_{\infty,2}(A, E)$ is a solid subspace of (E^A, C^A). If (E, C) is a Riesz space, then so is $(m_{\infty,2}(A, E), m^+_{\infty,2}(A, E))$. If C is $\mathcal{P}$-closed, then

$$m^+_\infty(A, E) \subset m^+_{\infty,2}(A, E) \quad \text{and} \quad m^+_\infty(A, E) = m^+_{\infty,2}(A, E) \cap m_\infty(A, E) ,$$

hence $m_\infty(A, E)$ is a solid subspace of $(m_{\infty,2}(A, E), m^+_{\infty,2}(A, E))$. Furthermore, if C is $\mathcal{P}$-closed and generating, then $E^{(A)} \subset m_{\infty,2}(A, E)$.

If q is a continuous monotone seminorm on E, then the functional λ''_q, defined by

$\lambda''_q(u_\iota, A) = \inf\{q''(u'') : u'' \in C'' , o \leq u_\iota \leq u'' \text{ in } E''\ \forall \iota \in A\}$ for all $(u_\iota, A) \in m^+_{\infty,2}(A, E)$, is a monotone sublinear functional on $m^+_{\infty,2}(A, E)$. According to (1.2.6), λ''_q induces the following monotone seminorms on $m_{\infty,2}(A, E)$:

(a)* $q''_\infty(x_i, A) = \inf\{\lambda''_q(u_i, A) + \lambda''_q(w_i, A) : (u_i, A), (w_i, A) \in m^+_{\infty,2}(A, E),$
$(u_i, A) - (w_i, A) = (x_i, A)\}$

(b)* $q''_{\infty S}(x_i, A) = \inf\{\lambda''_q(u_i, A) : (u_i, A) \in m^+_{\infty,2}(A, E),$
$(u_i, A) \pm (x_i, A) \in m^+_{\infty,2}(A, E)\}$

(c)* $q''_{\infty B}(x_i, A) = \inf\{\max(\lambda''_q(u_i, A), \lambda''_q(w_i, A)) : (u_i, A), (w_i, A) \in m^+_{\infty,2}(A, E),$
$-(u_i, A) \leq (x_i, A) \leq (w_i, A)\}$.

Furthermore, we have $q''_\infty = q''_{\infty D}$ and

$$q''_\infty = q''_{\infty S} = q''_{\infty B} = \lambda''_q \quad \text{on} \quad m^+_{\infty,2}(A, E) .$$

(2.2.9) Lemma. Let $(E, C, \mathcal{P})$ be an ordered convex space, let C be $\mathcal{P}$-closed and let q be a continuous monotone seminorm on E. Then the following statements hold:

(1) $\lambda''_q(J_\alpha(u)) = \lambda''_q(J_\infty(u)) = q(u)$ for all $u \in C$ and $\alpha \in \mathfrak{F}(A)$.

(2) $\sup q(u_j) \leq \lambda''_q(u_i, A)$ for all $(u_i, A) \in m^+_{\infty,2}(A, E)$.

(3) On $m_\infty(A, E)$ we have

$q''_\infty \leq q_\infty$, $q''_{\infty S} \leq q_{\infty S}$ and $q''_{\infty B} \leq q_{\infty B}$.

(4) If C is generating, then we have, for any $x \in E$, that

(i) $q''_\infty(J_\alpha(x)) \leq q_D(x)$ and $q''_\infty(J_\infty(x)) \leq q_D(x)$;

(ii) $q''_{\infty S}(J_\alpha(x)) = q''_{\infty S}(J_\infty(x)) = q_S(x)$;

(iii) $q''_{\infty B}(J_\alpha(x)) = q''_{\infty B}(J_\infty(x)) = q_B(x)$.

Proof. Conclusions (1) and (2) follow from the fact that $q'' = q$ on E and that q'' is monotone, while (3) is obvious in view of the definitions of q_∞ , $q_{\infty S}$ and $q_{\infty B}$. According to the conclusion (1) and the definitions of $q''_{\infty S}$ and $q''_{\infty B}$, it is easy to verify (4).

Let $(E, C, \mathcal{P})$ be a locally o-convex space and let $\mathcal{P}$ be determined by a family P of monotone seminorms. Denote by $\mathcal{P}_{\infty,2}$ the topology on $m_{\infty,2}(A, E)$ determined by $\{q''_{\infty} : q \in P\}$. Then $(m_{\infty,2}(A, E), m^+_{\infty,2}(A, E), \mathcal{P}_{\infty,2})$ is a locally solid space and $\mathcal{P}_{\infty,2}$ is also determined by $\{q''_{\infty S} : q \in P\}$ as well as by $\{q''_{\infty B} : q \in P\}$. If $(E, C, \mathcal{P})$ is metrizable or normable, then so is $(m_{\infty,2}(A, E), m^+_{\infty,2}(A, E), \mathcal{P}_{\infty,2})$. If C is $\mathcal{P}$-<u>closed</u>, then $m^+_{\infty,2}(A, E)$ is $\mathcal{P}_{\infty,2}$-closed, and the canonical embedding map from $m_{\infty}(A, E)$ into $m_{\infty,2}(A, E)$ is a continuous linear map from $(m_{\infty}(A, E), m^+_{\infty}(A, E), \mathcal{P}_{\infty})$ into $(m_{\infty,2}(A, E), m^+_{\infty,2}(A, E), \mathcal{P}_{\infty,2})$. If $(E, C, \mathcal{P})$ is a Fréchet space and if C is closed, then $(m_{\infty,2}(A, E), m^+_{\infty,2}(A, E), \mathcal{P}_{\infty,2})$ is a Fréchet locally solid space. If $(E, C, \mathcal{P})$ is a locally convex Riesz space, then so is $(m_{\infty,2}(A, E), m^+_{\infty,2}(A, E), \mathcal{P}_{\infty,2})$, and if, in addition, C is closed, then $m_{\infty}(A, E) = E^A \cap m_{\infty,2}(A, E)$.

Let $(E, C, \mathcal{P})$ be a locally o-convex space and let C be $\mathcal{P}$-<u>closed</u> and generating. Denote by $m_{0,2}(A, E)$ the $\mathcal{P}_{\infty,2}$-closure of $E^{(A)}$ in $m_{\infty,2}(A, E)$, by $\mathcal{P}_{\infty,o,2}$ the relative topology on $m_{0,2}(A, E)$ induced by $\mathcal{P}_{\infty,2}$, and by $m^+_{0,2}(A, E)$ the relative cone induced by $m^+_{\infty,2}(A, E)$, i.e.,

$$m^+_{0,2}(A, E) = m^+_{\infty,2}(A, E) \cap m_{0,2}(A, E) .$$

Then $(m_{0,2}(A, E), m^+_{0,2}(A, E), \mathcal{P}_{\infty,o,2})$ is a locally o-convex space, which is called the <u>space of almost-null-majorized families</u>.

It is known that the relative topology on $m_{\infty}(A, E)$ induced by $\mathcal{P}_{\infty,2}$ is coarser than $\mathcal{P}_{\infty}$, hence $m_0(A, E) \subset m_{0,2}(A, E)$ and the canonical embedding map is continuous from $(m_0(A, E), \mathcal{P}_{\infty,o})$ into $(m_{0,2}(A, E), \mathcal{P}_{\infty,o,2})$.

On the other hand, since $E^{(A)} \subset m_o(A, E) \subset m_{o,2}(A, E)$, it follows that $m_{o,2}(A, E)$ is the $\mathcal{P}_{\infty,2}$-closure of $m_o(A, E)$ in $m_{\infty,2}(A, E)$, thus $m_o(A, E)$ is dense in $m_{o,2}(A, E)$. Clearly $m^+_{o,2}(A, E)$ is $\mathcal{P}_{\infty,o,2}$-closed. It should be noted that $m^+_{o,2}(A, E)$ may not be generating. If $(E, C, \mathcal{P})$ is a locally convex Riesz space, then $m_{o,2}(A, E)$ is a solid subspace of $(m_{\infty,2}(A, E), m^+_{\infty,2}(A, E))$, and hence $(m_{o,2}(A, E), m^+_{o,2}(A, E), \mathcal{P}_{\infty,o,2})$ is a locally convex Riesz space.

Finally, using the same argument we are able to construct another typed space consisting of families. Let $(E, C, \mathcal{P})$ be an ordered convex space and let us define

$$p^+_o(A, E) = \{(u_i , A) \in C^A : \exists\, u \in C \text{ such that } \textstyle\sum_{i\in\alpha} u_i \leq u \ \ \forall\, \alpha \in \mathcal{F}(A)\}$$

$$p_o(A, E) = p^+_o(A, E) - p^+_o(A, E) .$$

Then $p^+_o(A, E)$ is a subcone of C^A , hence $p_o(A, E)$ is a solid subspace of (E^A, C^A) . If (E, C) is a Riesz space, then so is $(p_o(A, E), p^+_o(A, E))$. If C is generating, then $E^{(A)} \subset p_o(A, E)$. If $(E, C, \mathcal{P})$ is a locally o-convex space, then order-bounded subsets of E are $\mathcal{P}$-bounded, hence

$$p^+_o(A, E) \subset C_\omega(A, E) \quad\text{and}\quad p^+_o(A, E) = C_\omega(A, E) \cap p_o(A, E) ,$$

consequently, $p_o(A, E)$ is a solid subspace of $(\ell^1_\omega\langle A, E\rangle, C_\omega(A, E))$. If $(E, C, \mathcal{P})$ is quasi-complete and if C is $\mathcal{P}$-closed, then

$$C_\varepsilon(A, E) \subset p^+_o(A, E) \quad\text{and}\quad C_\varepsilon(A, E) = p^+_o(A, E) \cap \ell^1\langle A, E\rangle ,$$

and thus $\ell^1\langle A, E\rangle$ is a solid subspace of $(p_o(A, E), p^+_o(A, E))$.

If q is continuous monotone seminorm on E , Then the functional $q_{\omega,o}$ defined by

$$q_{\omega,o}(u_i , A) = \inf\{q(u) : u \in C , \sum_{i\in\alpha} u_i \leq u \ \ \forall \alpha \in \mathcal{F}(A)\}$$

$$(u_i , A) \in p_o^+(A, E)$$

is a monotone sublinear functional on $p_o^+(A, E)$ such that

$$\sup\{q(\Sigma_{i\in\alpha} u_i) : \alpha \in \mathcal{F}(A)\} \leq q_{\omega,o}(u_i , A) \ \ \forall (u_i , A) \in p_o^+(A, E). \qquad (2.10)$$

$q_{\omega,o}$ induces the following monotone seminorms on $p_o(A, E)$

(a) $\tilde{q}(x_i , A) = \inf\{q_{\omega,o}(u_i , A) + q_{\omega,o}(v_i , A) : (u_i , A), (v_i , A) \in p_o^+(A, E), (u_i , A) - (v_i , A) = (x_i , A)\}$

(b) $\tilde{q}_S(x_i , A) = \inf\{q_{\omega,o}(u_i , A) : (u_i , A) \in p_o^+(A, E), (u_i , A) \pm (x_i , A) \in p_o^+(A, E)\}$

(c) $\tilde{q}_B(x_i , A) = \inf\{\max(q_{\omega,o}(u_i , A), q_{\omega,o}(v_i , A)) : (u_i , A), (v_i , A) \in p_o^+(A, E), -(u_i , A) \leq (x_i , A) \leq (v_i , A)\}$.

Furthermore, we have $\tilde{q} = \tilde{q}_D$, and

$$\tilde{q} = \tilde{q}_S = \tilde{q}_B = q_{\omega,o} \quad \text{on} \quad p_o^+(A, E) .$$

(2.2.10) Lemma. <u>Let</u> $(E, C, \mathcal{P})$ <u>be a locally</u> o-<u>convex space and</u> q <u>a continuous monotone seminorm on</u> E . <u>Then the following statements hold</u>:

(1) $q_{\omega,o}(J_\alpha(u)) = kq(u)$, <u>where</u> $\alpha \in \mathcal{F}(A)$ <u>and</u> k <u>is the number of</u> α .

(2) $q_{\omega,D} \leq \tilde{q}$, $q_{\omega S} \leq \tilde{q}_S$ <u>and</u> $q_{\omega,B} \leq \tilde{q}_B$ <u>on</u> $p_o(A, E)$.

(3) <u>If</u> C <u>is generating, then we have, for any</u> $x \in E$, <u>that</u>

(i) $q_D(x) \leq \tilde{q}(J_\alpha(x)) \leq k\, q_D(x)$,

(ii) $q_S(x) \leq \tilde{q}_S(J_\alpha(x)) \leq k\, q_S(x)$,

(iii) $q_B(x) \leq \tilde{q}_B(J_\alpha(x)) \leq k\, q_B(x)$,

<u>where</u> $\alpha \in \mathcal{F}(A)$ <u>and</u> k <u>is the number of</u> α .

<u>Proof</u>. (1) follows from the monotonic property of q , (2) is a consequence of (2.10), while (3) is obvious.

Let $(E, C, \mathcal{P})$ be a locally o-convex space and let $\mathcal{P}$ be determined by a family P of monotone seminorms. Denote by $\mathcal{P}_o$ the topology on $p_0(A, E)$ determined by $\{\tilde{q} : q \in P\}$. Then $(p_0(A, E), p_0^+(A, E), \mathcal{P}_o)$ is a locally solid space, and $\mathcal{P}_o$ is determined by $\{\tilde{q}_S : q \in P\}$ as well as by $\{\tilde{q}_B : q \in P\}$. If $(E, C, \mathcal{P})$ is metrizable or normable, then so is $(p_0(A, E), p_0^+(A, E), \mathcal{P}_o)$. If C is closed, then so is $p_0^+(A, E)$. If $(E, C, \mathcal{P})$ is a Fréchet locally solid space and if C is closed, then $(p_0(A, E), p_0^+(A, E), \mathcal{P}_o)$ is a Fréchet locally solid space. If $(E, C, \mathcal{P})$ is a locally convex Riesz space, then so is $(p_0(A, E), p_0^+(A, E), \mathcal{P}_o)$.

It is worthwhile to notice that if (E, C) is an order complete Riesz space, then all ordered vector spaces constructed in this section are order complete Riesz spaces because they are all solid subspaces of the order complete Riesz space (E^A, C^A) .

It is known from (2.2.10) that the canonical embedding map from $p_0(A, E)$ into $\ell^1_\omega\langle A, E\rangle$ is continuous from $(p_0(A, E), p_0^+(A, E), \mathcal{P}_o)$ into $(\ell^1_\omega\langle A, E\rangle, C_\omega(A, E), \mathcal{P}_{\omega D})$, and that if $(E, C, \mathcal{P})$ is quasi-complete and if C is closed, then $\ell^1\langle A, E\rangle$ is a solid subspace of $(p_0(A, E), p_0^+(A, E))$. We shall prove a little more general result as follows.

(2.2.11) Lemma. <u>Let</u> $(E, C, \mathcal{P})$ <u>be a locally</u> o-<u>convex space</u>, <u>let</u> C <u>be closed and generating</u>, <u>and let</u> $\mathfrak{S}$ <u>be a family consisting of bounded subsets of</u> C <u>which satisfies the following two properties</u>:

(i) $\cup\{B : B \in \mathfrak{B}\}$ <u>generates a</u> $\mathcal{P}$-<u>dense subspace of</u> E ,

(ii) <u>if</u> $f \in C^*$ <u>is bounded on</u> B <u>for every</u> $B \in \mathfrak{B}$, <u>then</u> $f \in E'$.

<u>Denote by</u> $\mathcal{P}'$ <u>the</u> $\mathfrak{B}$-<u>topology on</u> E' . <u>Then the following statements hold:</u>

(1) $(E', C', \mathcal{P}')$ <u>is a locally o-convex space and</u> C' <u>is</u> $\mathcal{P}'$-<u>closed.</u>

(2) $\ell^1_\omega\langle A, E'\rangle = p_0(A, E')$ <u>and the equality is topological for the topology induced by</u> $\mathcal{P}'$.

(3) <u>If</u> $\mathcal{P}'$ <u>is consistent with</u> $\langle E, E'\rangle$, <u>then we also have that</u> $\ell^1\langle A, E'\rangle = p_0(A, E')$.

<u>Consequently,</u> $\ell^1\langle A, E'\rangle = p_0(A, E') = \ell^1_\omega\langle A, E\rangle$ <u>and the equalities are topological for the topology induced by</u> $\mathcal{P}'$.

<u>Proof</u>. (1) Clearly the topology $\mathcal{P}'$ is determined by a family $\{r_B : B \in \mathfrak{B}\}$ of seminorms, where each r_B is defined by

$$r_B(x') = \sup\{|\langle x, x'\rangle| : x \in B\} .$$

As $B \subset C$, it follows that r_B is monotone, hence $\mathcal{P}'$ is a locally o-convex topology.

The condition (i) obviously insures that $\sigma(E', E)$ is coarser than $\mathcal{P}'$, hence C' is $\mathcal{P}'$-closed.

(2) As order-bounded subsets of E' are $\mathcal{P}'$-bounded, it follows that $p_0(A, E') \subseteq \ell^1_\omega\langle A, E'\rangle$. On the other hand, if $(u_i', A) \in C_\omega(A, E')$, then for any $x \in E$, $(\langle x, u_i'\rangle, A)$ is summable, thus the following equation

$$\langle x, u'\rangle = \Sigma_A\langle x, u_i'\rangle \quad \text{for all} \quad x \in E \tag{2.12}$$

defines a positive linear functional u' on E with the following properties:

(a) u' is bounded on every $B \in \mathfrak{B}$ because the set of partial sums of (u_i', A) is $\mathcal{P}'$-bounded;

(b) $\Sigma_{i\in\alpha} u_i' \leq u'$ for all $\alpha \in \mathcal{F}(A)$;

(c) if $f \in C'$ is such that $\sum_{i\in\alpha} u_i' \leq f \ (\alpha \in \mathcal{F}(A))$, then $u' \leq f$.

According to (ii), (a) implies that $u' \in E'$. It then follows from (b) that $(u_i', A) \in p_0(A, E')$. Therefore we obtain $p_0(A, E') = \ell^1_\omega\langle A, E'\rangle$.

We are now going to verify that the equality is topological for the topology induced by $\mathcal{P}'$.

In fact, for any $B \in \mathfrak{B}$, we write $q = r_B$ where r_B is defined in (2.11). Then (c), (2.12) and the definition of $q_{\omega,o}$ imply that

$$q_{\omega,o}(u_i', A) = q(u') \quad \text{for all} \quad (u_i', A) \in C_\omega(A, E') .$$

On the other hand, by (2.2.1) and (2.11), we have

$$q_\omega(u_i', A) = \sup\{\sum_A \langle u, u_i'\rangle : u \in B\} = \sup\{\langle u, u'\rangle : u \in B\} = q(u') .$$

Therefore $q_{\omega,o} = q_\omega$, consequently $q_{\omega\, D} = \tilde{q}$, $q_{\omega\, S} = \tilde{q}_S$ and $q_{\omega\, B} = \tilde{q}_B$.

(3) As $\ell^1\langle A, E\rangle \subseteq \ell^1_\omega\langle A, E\rangle$, it follows from (2) that $\ell^1\langle A, E'\rangle \subseteq p_0(A, E')$. If $(u_i', A) \in C_\omega(A, E')$, then the net $\{\sum_{i\in\alpha} u_i' : \alpha \in \mathcal{F}(A)\}$ is directed upwards. The formula (2.12) shows that u' is the $\sigma(E', E)$-limit of $\{\sum_{i\in\alpha} u_i' : \alpha \in \mathcal{F}(A)\}$, and hence u' is the $\mathcal{P}'$-limit of $\{\sum_{i\in\alpha} u_i' : \alpha \in \mathcal{F}(A)\}$ because $\mathcal{P}'$ is locally o-convex and consistent with $\langle E, E'\rangle$ (see Wong and Ng [1, p.52]) . Consequently $\{\sum_{i\in\alpha} u_i' : \alpha \in \mathcal{F}(A)\}$ is a $\mathcal{P}'$-Cauchy net, thus $(u_i', A) \in \ell^1\langle A, E'\rangle$.

Finally, since $\mathcal{P}'_{\varepsilon D}$ is the relative topology on $\ell^1\langle A, E'\rangle$ induced by $\mathcal{P}'_{\omega D}$, it follows that the equality $\ell^1\langle A, E'\rangle = p_0(A, E')$ is also topological for the topology induced by $\mathcal{P}'$.

(2.2.12) Corollary. *Let* $(E, C, \|\cdot\|)$ *be an ordered Banach space with the norm* $\|\cdot\|$ *is monotone and let* C *be closed and generating. Then* $(u_i, A) \in C^A$ *belongs to* $p_0^+(A, E'')$ *if and only if* $(u_i, A) \in C_\omega(A, E)$, *and*

$$\|(u_i, A)\|_\omega = \|(u_i, A)\|''_{\omega,o} \quad \text{for all } (u_i, A) \in C_\omega(A, E). \tag{2.13}$$

where $\|\cdot\|''_{\omega,o}$ *is the norm on* $p_0(A, E'')$ *induced by the bidual norm* $\|\cdot\|''$ *of* $\|\cdot\|$.

Proof. We first note that $(E, C, \|\cdot\|)$, $(E', C', \|\cdot\|')$ and $(E'', C'', \|\cdot\|'')$ are locally solid spaces. If $(u_i, A) \in C^A$ belongs to $p_0^+(A, E'')$, then the set of partial sums of (u_i, A) is topologically bounded in E'' and surely in E, thus $(u_i, A) \in C_\omega(A, E)$. Conversely, if $(u_i, A) \in C_\omega(A, E)$, then its partial sums are bounded in E'', hence $(u_i, A) \in C_\omega(A, E'')$. On the other hand, the norm-topology on E'' is an $\mathfrak{S}$-topology satisfying all conditions in (2.2.12), hence $C_\omega(A, E'') = p_0^+(A, E'')$ by (2.2.11) ; therefore $(u_i, A) \in C^A$ belongs to $p_0^+(A, E'')$. Finally, the equality (2.13) follows from $C_\omega(A, E'') = p_0^+(A, E'')$ and (2.2.11) on account of the fact that $\|\cdot\|$ is the restriction of $\|\cdot\|''$ on E.

Let $(E, C, \mathcal{P})$ be a locally o-convex space and let C be generating. We have known that $C^{(A)}$ is a generating cone in $E^{(A)}$, and that $E^{(A)}$ is a solid subspace of the locally solid spaces constructed in this section, except for $(m_{\infty,2}(A, E), m^+_{\infty,2}(A, E), \mathcal{P}_{\infty,2})$. If, in addition, C is closed, then $E^{(A)}$ is a solid subspace of $(m_{\infty,2}(A, E), m^+_{\infty,2}(A, E), \mathcal{P}_{\infty,2})$. Denote by $E_{\varepsilon,D}^{(A)}$ (resp. $E_{\infty}^{(A)}$ and $E_{o}^{(A)}$) the ordered vector space $(E^{(A)}, C^{(A)})$ equipped with the relative topology induced by $\mathcal{P}_{\varepsilon,D}$ (resp. $\mathcal{P}_{\infty}$ and $\mathcal{P}_{o}$). If C is $\mathcal{P}$-closed, then we denote by $E_{\infty,2}^{(A)}$ the space $(E^{(A)}, C^{(A)})$ equipped with the relative topology induced by $\mathcal{P}_{\infty,2}$. All these spaces are locally solid as the following result shows.

(2.2.13) Lemma. <u>Let</u> $(E, C, \mathcal{P})$ <u>be a locally o-convex space and</u> <u>let</u> C <u>be generating. Then</u> $E_{\varepsilon,D}^{(A)}$, $E_{\infty}^{(A)}$ <u>and</u> $E_{o}^{(A)}$ <u>are locally solid spaces.</u> <u>Furthermore, if in addition,</u> C <u>is</u> $\mathcal{P}$-<u>closed, then</u> $E_{\infty,2}^{(A)}$ <u>is a locally solid space.</u>

<u>Proof</u>. We only give the proof for $E_{\varepsilon,D}^{(A)}$ because the proofs of other cases are similar. Clearly $E_{\varepsilon,D}^{(A)}$ is a locally o-convex space because subspaces of a locally o-convex space are locally o-convex. Let $\mathcal{P}$ be determined by a family P of monotone seminorms. Then $\{q_{\varepsilon S} : q \in P\}$ determines the locally solid topology $\mathcal{P}_{\varepsilon,D}$, and each $q_{\varepsilon S}$ is a Riesz seminorm on $(\ell^1 \langle A, E \rangle, C_{\varepsilon}(A, E))$. In order to verify that $E_{\varepsilon,D}^{(A)}$ be locally decomposable, it is sufficient to show that for any $(x_\iota, (A)) \in E^{(A)}$ and $\delta > 0$, there exists $(w_\iota, (A)) \in C^{(A)}$ with $-(w_\iota, (A)) \leqslant (x_\iota, (A)) \leqslant (w_\iota, (A))$ such that

$$q_{\varepsilon}(w_\iota, (A)) < q_{\varepsilon S}(x_\iota, (A)) + \delta .$$

In fact, we first note that there exists $(u_\iota, A) \in C_{\omega}(A, E)$ with $-(u_\iota, A) \leqslant (x_\iota, (A)) \leqslant (u_\iota, A)$ such that

$$q_{\varepsilon}(u_\iota, A) < q_{\varepsilon S}(x_\iota, (A)) + \delta \qquad (2.14)$$

since $q_{\varepsilon S}$ is a Riesz seminorm on $\ell^1 \langle A, E \rangle$. As $(x_\iota, (A)) \in E^{(A)}$, we may define, for any $\iota \in A$, that

$$w_\iota = \begin{cases} u_\iota & \text{if } x_\iota \neq 0 \\ o & \text{if } x_\iota = 0 . \end{cases}$$

Then $(w_i, A) \in C^{(A)}$ is such that $-(w_i, A) \leq (x_i, (A)) \leq (w_i, A)$. Clearly $0 \leq (w_i, A) \leq (u_i, A)$.

As q_ε is monotone, it follows from (2.14) that

$$q_\varepsilon(w_i, A) \leq q_\varepsilon(u_i, A) < q_{\varepsilon S}(x_i, (A)) + \delta$$

which obtains our required assertion.

(2.2.14) Corollary. <u>Let</u> $(E, C, \mathcal{P})$ <u>be a Fréchet locally solid space and let</u> C <u>be closed</u>. <u>Then</u> $(m_0(A, E), m_0^+(A, E), \mathcal{P}_{\infty\, 0})$ <u>and</u> $(m_{0,2}(A, E), m_{0,2}^+(A, E), \mathcal{P}_{\infty, 0, 2})$ <u>are Fréchet locally solid spaces</u>.

<u>Proof</u>. It is known that $(m_\infty(A, E), m_\infty^+(A, E), \mathcal{P}_\infty)$ and $(m_{\infty,2}(A, E), m_{\infty,2}^+(A, E), \mathcal{P}_{\infty,2})$ are Fréchet locally solid spaces, hence $(m_0(A, E), \mathcal{P}_{\infty\, 0})$ is the completion of $E_\infty^{(A)}$ and $(m_{0,2}(A, E), \mathcal{P}_{\infty,2})$ is the completion of $E_{\infty,2}^{(A)}$ by the definitions of $m_0(A, E)$ and $m_{0,2}(A, E)$. On the other hand, the $\mathcal{P}_\infty$-closure of $C^{(A)}$ in $m_\infty(A, E)$ is contained in $m_0^+(A, E)$ since $m_0^+(A, E)$ is $\mathcal{P}_\infty$-closed, and the $\mathcal{P}_{\infty,2}$-closure of $C^{(A)}$ in $m_{\infty,2}(A, E)$ is contained in $m_{0,2}^+(A, E)$. By making use of (1.3.7), $(m_0(A, E), m_0^+(A, E), \mathcal{P}_{\infty\, 0})$ and $(m_{0,2}(A , E), m_{0,2}^+(A, E), \mathcal{P}_{\infty,0,2})$ are Fréchet locally solid spaces.

The following result is concerned with the continuity of the identity map on $E^{(A)}$.

(2.2.15) Lemma. <u>Let</u> $(E, C, \mathcal{P})$ <u>be a locally</u> o-<u>convex space, let</u> C <u>be generating and let</u> q <u>be a continuous monotone seminorm on</u> E. <u>Then the following statements hold</u>:

(1) $q_\infty \leq q_\varepsilon$ <u>on</u> $C^{(A)}$, <u>hence the identity map</u> $E_\varepsilon^{(A)} \to E_\infty^{(A)}$ <u>is continuous on</u> $C^{(A)}$.

(2) $q_{(n)} \leq q_\infty \leq q_{\varepsilon D} \leq \tilde{q}$ <u>on</u> $E^{(A)}$, <u>hence the identity maps</u>

$$E_o^{(A)} \to E_{\varepsilon,D}^{(A)} \to E_\infty^{(A)} \to E_{(n)}^{(A)}$$

<u>are all continuous</u>.

(3) <u>If</u> C <u>is</u> $\mathcal{P}$<u>-closed, then</u>

$$q_{(n)} \leq q''_\infty \leq q_\infty \quad \text{on} \quad E^{(A)},$$

<u>consequently, the identity maps</u>

$$E_\infty^{(A)} \to E_{\infty,2}^{(A)} \to E_{(n)}^{(A)}$$

<u>are continuous</u>.

<u>Proof</u>. In the following proof, we assume that V is the unit ball of q in E. If $(x_i, (A)) \in E^{(A)}$ (or $C^{(A)}$), then there exists $\alpha \in \mathcal{F}(A)$ such that

$$x_j = 0 \text{ for all } j \notin \alpha .$$

(1) Let $(u_i, (A)) \in C^{(A)}$ and let $\alpha \in \mathcal{F}(A)$ be such that $u_j = 0 \ \forall j \notin \alpha$. Then $u = \sum_{j\in\alpha} u_j \in C$ is such that $o \leq u_i \leq u$ for all $i \in \alpha$ and

$$|\langle \sum_{j\in\alpha} u_j, f\rangle| \leq \sum_{j\in\alpha} |\langle u_j, f\rangle| \leq \sup\{\sum_{j\in\alpha} |\langle u_j, g\rangle| : g \in V^o\}$$
$$= q_\varepsilon(u_i, (A)) \quad \text{for all } f \in V^o ,$$

It follows from the definition λ_q that

$$\lambda_q(u_i, (A)) \leq q(u) \leq q_\varepsilon(u_i, (A)) .$$

As $q_\infty = \lambda_q$ on $m_\infty^+(A, E)$, we obtain

$$q_\infty(u_\iota, (A)) \leq q_\varepsilon(u_\iota, (A)) .$$

(2) As $q_\infty = \lambda_q$ on $m_\infty^+(A, E)$ and $q_{\varepsilon D} = q_\varepsilon$ on $C_\varepsilon(A, E)$, the inequality $q_\infty \leq q_{\varepsilon D}$ is an immediate consequence of the conclusion (1) and the definitions of q_∞ and $q_{\varepsilon D}$. While the inequality $q_{\varepsilon D} \leq \tilde{q}$ is a consequence of (2.2.10)(2) according to $q_{\varepsilon D} = q_{\omega D}$ on $\ell^1\langle A, E\rangle$. To prove $q_{(n)} \leq q_\infty$, let $(x_\iota, (A)) \in E^{(A)}$ and let $(u_\iota, (A))$ and $(w_\iota, (A))$, in $C^{(A)}$, be such that $(x_\iota, (A)) = (u_\iota, (A)) - (w_\iota, (A))$. Then we have

$$q(x_\iota) \leq q(u_\iota) + q(w_\iota) \leq \sup\{q(u_\iota) + \sup q(w_\iota) \leq \lambda_q(u_\iota, (A)) + \lambda_q(w_\iota, (A))$$
$$\forall \iota \in A .$$

the last inequality follows from (2.2.9)(2), hence

$$q_{(n)}(x_\iota, (A)) = \sup q(x_\iota) \leq \lambda_q(u_\iota, (A)) + \lambda_q(w_\iota, (A)) .$$

We conclude from the definition of q_∞ that

$$q_{(n)}(x_\iota, (A)) \leq q_\infty(x_\iota, (A)) .$$

(3) The inequality $q''_\infty \leq q_\infty$ follows from (2.2.9)(3), while the proof of $q_{(n)} \leq q''_\infty$ is almost the same as that of $q_{(n)} \leq q_\infty$ when λ_q is instead of λ''_q.

We shall see in (2.4.11) that $q''_{\infty B} = q_{\infty B}$ on $E^{(A)}$, therefore the relative topology on $E^{(A)}$ induced by $\mathcal{P}_{\infty,2}$ coincides with the relative topology induced by $\mathcal{P}_\infty$.

2.3 The topological dual of $\ell^1\langle A, E\rangle$

Let $(E, C, \mathcal{P})$ be an ordered convex space with the topological dual E'. If G and H are respectively vector subspaces of E^A and $(E')^A$ such that $(\langle x_i, x_i'\rangle, A)$ are summable families in $\mathbb{R}$ for all $(x_i, A) \in G$ and $(x_i', A) \in H$, then one can define a bilinear form on $G \times H$ by setting

$$\langle (x_i, A), (x_i', A)\rangle = \sum_A \langle x_i, x_i'\rangle \tag{3.1}$$

If the bilinear form (3.1) satisfies the following condition

$$\langle (x_i, A), (x_i', A)\rangle = 0 \quad \text{for all} \quad (x_i, A) \in G \Rightarrow (x_i', A) = 0, \tag{3.2}$$

then the bilinear form (3.1) induces an algebraic isomorphism from H into the algebraic dual G^* of G. Clearly, if $E^{(A)} \subset G$, then (3.2) is satisfied; if $(E')^{(A)} \subset H$, then the bilinear form (3.1) induces an algebraic isomorphism from G into the algebraic dual H^* of H. For instance, the following each pair insures that $(|\langle x_i, x_i'\rangle|, A)$ is summable:

(a) $(x_i, A) \in \ell^1[A, E]$ and (x_i', A) is an equicontinuous family;

(b) $(x_i, A) \in \ell^1(A, E)$ and (x_i', A) is a prenuclear family;

(c) $(x_i, A) \in c_o(A, E)$ and $\sum_A p_B(x_i') < \infty$, where B and p_B are defined as in (2.1.5).

If $E^{(A)} \subset G$, as $(G, G \cap C^A)$ and $(H, H \cap (C')^A)$ are ordered vector spaces, it follows that the algebraic isomorphism from H into G^*, induced by (3.1), preserves the ordering.

Let $(E, C, \mathcal{P})$ be a locally o-convex space and let C be generating. Then $E^{(A)} \subset \ell^1\langle A, E\rangle \subset \ell^1_\omega\langle A, E\rangle$ and $(E')^{(A)} \subset m_\infty(A, E')$. If

$(u_\iota, A) \in C_\omega(A, E)$ and if $(u'_\iota, A) \in m^+_\infty(A, E')$, then there exists $u' \in C'$ such that $o \leq u'_\iota \leq u'$ for all $\iota \in A$, hence

$$\sum_A \langle u_\iota, u'_\iota\rangle \leq \sum_A \langle u_\iota, u'\rangle < \infty$$

As $C_\omega(A, E) = \ell^1_\omega\langle A, E\rangle \cap C^A$ and $m^+_\infty(A, E') = m_\infty(A, E') \cap (C')^A$ are generating cones in $\ell^1_\omega\langle A, E\rangle$ and $m_\infty(A, E')$ respectively, it follows that $\langle \ell^1_\omega\langle A, E\rangle, m_\infty(A, E')\rangle$ is a dual pair under the bilinear form (3.1). Therefore, $m_\infty(A, E')$ can be regarded as a vector subspace of the algebraic dual of $\ell^1_\omega\langle A, E\rangle$, and hence elements in $m^+_\infty(A, E')$ are positive linear functionals on $(\ell^1_\omega\langle A, E\rangle, C_\omega(A, E))$. Furthermore, elements in $m_\infty(A, E')$ are $\mathcal{P}_{\omega D}$-continuous as the following result shows.

(2.3.1) Lemma. Let $(E, C, \mathcal{P})$ be a locally o-convex space and let C be generating. Then $m_\infty(A, E')$ may be identified with a subspace of $(\ell^1_\omega\langle A, E\rangle, C_\omega(A, E), \mathcal{P}_{\omega D})'$, and hence elements in $m^+_\infty(A, E')$ are positive $\mathcal{P}_{\omega D}$-continuous linear functional on $\ell^1_\omega\langle A, E\rangle$.

Proof. As $m^+_\infty(A, E')$ is generating, it is sufficient to show that elements in $m^+_\infty(A, E')$ are $\mathcal{P}_{\omega D}$-continuous. Let $(u'_\iota, A) \in m^+_\infty(A, E')$ and let u', in C', be such that $o \leq u'_\iota \leq u'$ for all $\iota \in A$. Then there exists a continuous monotone seminorm q on E such that

$$|\langle x, u'\rangle| \leq q(x) \quad \text{for all} \quad x \in E .$$

The $\mathcal{P}_{\omega D}$-continuity of (u'_ι, A) will follow by showing that

$$\left|\sum_A \langle x_\iota, u'_\iota\rangle\right| \leq q_{\omega D}(x_\iota, A) \quad \text{for all} \quad (x_\iota, A) \in \ell^1_\omega\langle A, E\rangle. \tag{3.3}$$

In fact, for any $(x_i, A) \in \ell^1_\omega\langle A, E\rangle$ and $\delta > 0$, there exist (v_i, A) and (w_i, A) in $C_\omega(A, E)$ with $(x_i, A) = (v_i, A) - (w_i, A)$ such that

$$q_\omega(v_i, A) + q_\omega(w_i, A) < q_{\omega D}(x_i, A) + \delta .$$

By (2.2.1), we have, for any $\alpha \in \mathcal{F}(A)$, that

$$q(\sum_{i\in\alpha} v_i) + q(\sum_{i\in\alpha} w_i) < q_{\omega D}(x_i, A) + \delta .$$

It then follows that

$$\begin{aligned}\sum_{i\in\alpha}|\langle x_i, u'_i\rangle| &\leq \sum_{i\in\alpha}\langle v_i, u'_i\rangle + \sum_{i\in\alpha}\langle w_i, u'_i\rangle \\ &\leq \sum_{i\in\alpha}\langle v_i, u'\rangle + \sum_{i\in\alpha}\langle w_i, u'\rangle \\ &\leq q(\sum_{i\in\alpha} v_i) + q(\sum_{i\in\alpha} w_i) < q_{\omega D}(x_i, A) + \delta .\end{aligned}$$

and hence that

$$|\sum_A \langle x_i, u'_i\rangle| \leq \sum_A |\langle x_i, u'_i\rangle| \leq q_{\omega D}(x_i, A) + \delta .$$

As δ was arbitrary, we get the required inequality (3.3).

(2.3.2) Lemma. <u>Let</u> $(E, C, \mathcal{P})$ <u>be a locally solid space</u>. <u>There exists a positive projection</u> $\prod$ <u>from</u> $(\ell^1_\omega\langle A, E\rangle, C_\omega(A, E), \mathcal{P}_{\omega D})'$ <u>onto</u> $m_\infty(A, E')$ <u>such that</u>

$$\ker \prod = (\ell^1\langle A, E\rangle)^\perp \quad \underline{\text{and}} \quad \prod((C_\omega(A, E))') = m^+_\infty(A, E') ,$$

<u>where</u> $(\ell^1\langle A, E\rangle)^\perp$ <u>is the polar of</u> $\ell^1\langle A, E\rangle$ <u>taken in</u> $(\ell^1_\omega\langle A, E\rangle, C_\omega(A, E), \mathcal{P}_{\omega D})'$.

<u>Proof</u>. For each $\iota \in A$, as $(E, C, \mathcal{P})$ is locally solid and surely <u>locally decomposable</u>, it follows from (2.2.6) that the natural map J_ι is a positive continuous (injective) linear map from $(E, C, \mathcal{P})$ into $(\ell^1_\omega\langle A, E\rangle, C_\omega(A, E), \mathcal{P}_{\omega D})$, and hence that $x'_\iota = f \circ J_\iota \in E'$ for any $\mathcal{P}_{\omega D}$-continuous linear functional f on $\ell^1_\omega\langle A, E\rangle$. Therefore we may define a map $\prod : (\ell^1_\omega\langle A, E\rangle, C_\omega(A, E), \mathcal{P}_{\omega D})' \to ((E')^A, (C')^A)$ by setting

$$\prod(f) = (f \circ J_\iota , A) \quad \text{for any} \quad f \in (\ell^1_\omega\langle A, E\rangle, C_\omega(A, E), \mathcal{P}_{\omega D})' . \tag{3.4}$$

clearly $\prod$ is linear. As J_ι is positive, it follows that $f \circ J_\iota \in C'$ $(\iota \in A)$ whenever $f \in C_\omega(A, E)'$, and hence that $\prod$ is a positive linear map.

In order to verify that the range of $\prod$ is contained on $m_\infty(A, E')$, it is sufficient to show that $\prod(C_\omega(A, E)') \subset m^+_\infty(A, E')$ because $C_\omega(A, E)'$ is generating. Let $g \in C_\omega(A, E)'$ and let q be a continuous monotone seminorm on E such that

$$|\langle (x_\iota , A), g\rangle| \leq q_{\omega D}(x_\iota , A) \quad \text{for all} \quad (x_\iota , A) \in \ell^1_\omega\langle A, E\rangle . \tag{3.5}$$

Define, for any $u \in C$, that

$$r_g(u) = \sup\{\langle (u_\iota , A), g\rangle : (u_\iota , A) \in C^{(A)} \text{ with } \textstyle\sum_{\iota\in A} u_\iota = u\}. \tag{3.6}$$

Then r_g is a superlinear functional on C such that

$$\langle J_\iota(u), g\rangle \leq r_g(u) \quad \text{for all} \quad u \in C \quad \text{and} \quad \iota \in A \tag{3.7}$$

because of $\sum_{j\in A} \pi_j(J_\iota(u)) = u$, where π_j is the j-th projection from E^A into E . Moreover, for any $(u_\iota , A) \in C^{(A)}$ with $\sum_{\iota\in A} u_\iota = u$, we have, by (2.2.1), that

$$q_{\omega D}(u_\iota , A) = \sup\{q(\sum_{\iota\in\alpha} u_\iota) : \alpha \in \mathcal{F}(A)\} = q(u) \tag{3.8}$$

because q is monotone. On account of (3.5), (3.8) and (3.6), we obtain

$$r_g(u) \leqslant q(u) \quad \text{for all} \quad u \in C .$$

By Bonsall's theorem (1.1.1), there exists a positive linear functional u' on E such that

$$r_g(u) \leqslant \langle u, u'\rangle \quad (u \in C) \quad \text{and} \quad \langle x, u'\rangle \leqslant q(x) \quad (x \in E) , \tag{3.9}$$

hence $u' \in E'$ (in fact $u' \in V_q^o$) and

$$\langle u, g \circ J_\iota\rangle = \langle J_\iota(u), g\rangle \leqslant r_g(u) \leqslant \langle u, u'\rangle \quad (u \in C)$$

by making use of (3.7). The above inequalities show that

$$g \circ J_\iota \leqslant u' \quad \text{for all} \quad \iota \in A \tag{3.10}$$

or, equivalently $\prod(g) = (g \circ J_\iota , A) \in m_\infty^+(A, E')$ since $\prod$ is positive.

Next we show that $\prod(f) = f$ for all $f = (x_\iota', A) \in m_\infty(A, E')$. In view of the definition of $\prod$ (see (3.4)), we have that $\prod(f) = (f \circ J_\iota , A)$. On account of (3.1), we obtain

$$\langle x, f \circ J_\iota\rangle = \langle J_\iota(x), f\rangle = \langle x, x_\iota'\rangle \quad \text{for all} \quad x \in E ,$$

hence $f \circ J_\iota = x_\iota'$ for all $\iota \in A$, consequently, $\prod(f) = f$.

From the above two conclusions, $\prod$ is a positive projection from $(\ell_\omega^1\langle A, E\rangle, C_\omega(A, E), \mathcal{P}_{\omega D})'$ onto $m_\infty(A, E')$.

As $\prod$ is a positive projection and $m_\infty^+(A, E') \subseteq C_\omega(A, E)'$, we have

$$\prod(C_\omega(A, E)') \subseteq m_\infty^+(A, E') = \prod(m_\infty^+(A, E')) \subseteq \prod(C_\omega(A, E)')$$

which implies that $m_\infty^+(A, E') = \prod(C_\omega(A , E)')$.

Finally we show that $\ker \prod = (\ell^1\langle A, E\rangle)^\perp$. In fact

$$f \in \ker \prod \Leftrightarrow f \circ J_\iota = 0 \quad \text{on} \quad E \quad (\text{for all} \quad \iota \in A) \Leftrightarrow f = o \quad \text{on} \quad E^{(A)} .$$

As $\ell^1\langle A, E\rangle$ is the $\mathscr{P}_{\omega D}$-closure of $E^{(A)}$ in $\ell^1_\omega\langle A, E\rangle$ (see (2.2.7)), we have that $f = 0$ on $E^{(A)}$ if and only if $f = 0$ on $\ell^1\langle A, E\rangle$ on account of the continuity of f . Therefore $\ker \prod = (\ell^1\langle A, E\rangle)^\perp$.

(2.3.3) Theorem (Walsh [1]). <u>Let</u> $(E, C, \mathscr{P})$ <u>be a locally solid space</u>. <u>Then</u> $(\ell^1\langle A, E\rangle, C_\varepsilon(A, E), \mathscr{P}_{\varepsilon D})'$ <u>may be identified with</u> $m_\infty(A, E')$.

<u>Proof</u>. According to (2.3.2), there exists a positive projection $\prod$ from $(\ell^1_\omega\langle A, E\rangle)'$ onto $m_\infty(A, E')$ such that $\ker \prod = (\ell^1\langle A, E\rangle)^\perp$. Then the bijection $\hat{\prod}$ associated with $\prod$ is an algebraic isomorphism from $(\ell^1_\omega\langle A, E\rangle)' / (\ell^1\langle A, E\rangle)^\perp$ onto $m_\infty(A , E')$. Since $\mathscr{P}_{\omega D}$ is the relative topology induced by $\mathscr{P}_{\omega D}$, it follows from a well-known result that $(\ell^1\langle A, E\rangle)' = (\ell^1\langle A, E\rangle, \mathscr{P}_{\omega D})'$ and $(\ell^1_\omega\langle A, E\rangle)' / (\ell^1\langle A, E\rangle)^\perp$ are algebraically isomorphic. Therefore $(\ell^1\langle A, E\rangle)'$ and $m_\infty(A, E')$ are algebraically isomorphic.

(2.3.4) Proposition. <u>Let</u> $(E, C, \mathscr{P})$ <u>be a locally solid space and</u> $p(A, E')$ <u>the space consisting of all prenuclear families</u> (<u>with index set</u> A) <u>in</u> E' . <u>Then</u> $m_\infty(A, E')$ <u>is the order-convex hull of</u> $p(A, E')$ <u>in</u> $((\ell^1\langle A, E\rangle)^*, C_\varepsilon(A, E)^*)$.

<u>Proof</u>. In view of (2.1.4), $p(A, E')$ can be regarded as the topological dual $(\ell^1(A, E))'$ of $(\ell^1(A, E), \mathscr{P}_\varepsilon)$. Since $E^{(A)} \subset \ell^1\langle A, E\rangle \subset \ell^1(A, E)$ and since $E^{(A)} \subset \ell^1\langle A, E\rangle \subset \ell^1(A, E)$ and since $E^{(A)}$ is $\mathscr{P}_\varepsilon$-dense in $\ell^1(A, E)$, it follows that $p(A, E')$ can be regarded as the topological dual of $\ell^1\langle A, E\rangle$ equipped with the relative topology induced by $\mathscr{P}_\varepsilon$. Since $\mathscr{P}_{\varepsilon D}$ is the locally decomposable topology associated with the relative topology induced by $\mathscr{P}_\varepsilon$, and since $m_\infty(A, E')$ is the topological dual $(\ell^1\langle A, E\rangle)'$ of $(\ell^1\langle A, E\rangle, \mathscr{P}_{\varepsilon D})$, it follows from Wong and Ng [1, (3.12) p.38] that $m_\infty(A, E')$ is the order-convex hull of $p(A, E')$ in $(\ell^1\langle A, E\rangle)^*, C_\varepsilon(A, E)^*)$.

For a locally solid space $(E, C, \mathscr{P})$, we have known from (2.3.1) and (2.3.3) that $\langle \ell^1_\omega\langle A, E\rangle, m_\infty(A, E')\rangle$ and $\langle \ell^1\langle A, E\rangle, m_\infty(A, E')\rangle$ are dual pairs, and that $\mathscr{P}_{\varepsilon D}$ is a locally solid topology on $\ell^1\langle A, E\rangle$ which is consistent with $\langle \ell^1\langle A, E\rangle, m_\infty(A, E')\rangle$. For a given subset B of $\ell^1_\omega\langle A, E\rangle$, we denote by B^p the polar of B taken in $m_\infty(A, E')$, and by B^o the polar of B taken in the topological dual of $(\ell^1_\omega\langle A, E\rangle, C_\omega(A, E), \mathscr{P}_{\omega D})$.

(2.3.5) Theorem (Walsh [1]). <u>Let</u> $(E, C, \mathscr{P})$ <u>be a locally solid space, let</u> q <u>be a continuous monotone seminorm on</u> E, <u>and let</u> $\prod$ <u>be the positive projection from</u> $(\ell^1_\omega\langle A, E\rangle)'$ <u>onto</u> $m_\infty(A, E')$ <u>defined in</u> (2.3.2). <u>Suppose further that</u> V <u>is the unit ball of</u> q, <u>i.e.</u>, $V = \{x \in E : q(x) < 1\}$, <u>and that</u>

$$V_{\omega D} = \{(x_i, A) \in \ell^1_\omega\langle A, E\rangle : q_{\omega D}(x_i, A) < 1\} ;$$
$$V_{\omega S} = \{(x_i, A) \in \ell^1_\omega\langle A, E\rangle : q_{\omega S}(x_i, A) < 1\} .$$

<u>Then the following statements hold:</u>

(a) $\prod(V^o_{\omega D} \cap C_\omega(A\ , E)') = V^p_{\omega D} \cap m^+_\infty(A, E')$

$= \{(u'_i , A) \in m^+_\infty(A, E') : \exists\, u' \in V^o \cap C'$ <u>such that</u>

$o \leq u'_i \leq u' \quad \forall\, i \in A\}$.

(b) $\prod(V^o_{\omega D}) = V^p_{\omega D}$.

(c) $V^o_{\omega D} \cap C_\omega(A, E)' = V^o_{\omega S} \cap C_\omega(A, E)' = V^o_{\omega B} \cap C_\omega(A, E)'$

(d) $\prod(V^o_{\omega S}) = V^p_{\omega S}$

$= \{(x'_i , A) \in m_\infty(A, E') : \exists\, u' \in V^o \cap C'$ <u>such that</u> $-u' \leq x'_i \leq u'$

$\forall\, i \in A\}$.

<u>Proof</u>. For simplicity of notation, we write

$M^+_\infty = \{(u'_i , A) \in m^+_\infty(A, E') : \exists\, u' \in V^o \cap C'$ such that $o \leq u'_i \leq u' \ \forall\, i \in A\}$;

$M_{\infty S} = \{(x'_i , A) \in m_\infty(A, E') : \exists\, u' \in V^o \cap C'$ such that $-u' \leq u'_i \leq u' \ \forall\, i \in A\}$;

$V_\omega = \{(x_i , A) \in \ell^1_\omega\langle A, E\rangle : q_\omega(x_i , A) < 1\}$.

Then V_ω is positively order-convex because q_ω is monotone (see (2.2.1)), and

$$V_{\omega D} = D(V_\omega) = co(-(V_\omega \cap C_\omega(A, E)) \cup (V_\omega \cap C_\omega(A, E))) ,$$

$$V_{\omega S} = S(V_\omega)$$

in view of (1.2.5) and the definitions of $q_{\omega D}$ and $q_{\omega S}$. Also the positive order-convexity of V_ω insures that

$$V_\omega \cap C_\omega(A, E) = V_{\omega D} \cap C_\omega(A, E) = V_{\omega S} \cap C_\omega(A, E) . \qquad (3.11)$$

The proof is complete by considering the following steps:

(i) $M^+_\infty \subseteq V^p_{\omega D} \cap m^+_\infty(A, E')$.

Let $(u'_\iota, A) \in M^+_\infty$ and let u' , in $V^\circ \cap C'$, be such that $o \leq u'_\iota \leq u'$ $(\iota \in A)$. As $|\langle x, u'\rangle| \leq q(x)$ $(x \in E)$, it follows from (3.3) that $(u'_\iota, A) \in V^p_{\omega D} \cap m^+_\infty(A, E')$ as required.

(ii) $M^+_\infty = \prod(V^\circ_{\omega D} \cap C_\omega(A, E)')$.

Since $V^p_{\omega D} \subset V^\circ_{\omega D}$, $m^+_\infty(A, E') \subset C_\omega(A, E)'$ and since $\prod$ is a positive projection, it follows from the conclusion (i) that $M^+_\infty \subseteq \prod(V^\circ_{\omega D} \cap C_\omega(A, E)')$.

Conversely, if $g \in V^\circ_{\omega D} \cap C_\omega(A, E)'$, then by (3.9) and (3.10), there exists $u' \in V^\circ \cap C'$ such that $o \leq g \circ J_\iota \leq u'$ $(\iota \in A)$, and hence from the definition of $\prod$ that $\prod(g) = (g \circ J_\iota, A) \in M^+_\infty$, as required.

(iii) $M^+_\infty = V^p_{\omega D} \cap m^+_\infty(A, E')$.

This follows from the conclusion (ii) by making use the fact that $\prod$ is a projection and $V^p_{\omega D} \cap m^+_\infty(A , E') \subset V^\circ_{\omega D} \cap C_\omega(A, E)'$.

(iv) $V^\circ_{\omega D} = F(S(V^\circ_\omega)) = F(D(V^\circ_\omega)) = B(V^\circ_\omega)$.

Since $\mathscr{P}_{\omega D}$ is finer than the relative topology induced by $\mathscr{P}_\omega$, it follows that V_ω , $V_{\omega D}$ and $V_{\omega S}$ are $\mathscr{P}_{\omega D}$-neighbourhoods of 0 , and hence the conclusion follows from **Proposition (1.1.11)**.

(v) $\prod(V^\circ_{\omega D}) = V^p_{\omega D}$.

As $\prod$ is projection, it is clear that $V^p_{\omega D} \subseteq \prod(V^\circ_{\omega D})$. For any $f \in V^\circ_{\omega D}$, there exist, by (iv), $g, h \in V^\circ_\omega \cap C_\omega(A, E)'$ such that $-h \leq f \leq g$, hence

$$- \prod(h) \leq \prod(f) \leq \prod(g) .$$

For any $(x_i , A) \in V_{\omega,D}$, on account of $V_{\omega D} = D(V_\omega)$, there exist $\lambda \in (0, 1)$ and $(u_i , A), (w_i , A)$ in $V_\omega \cap C_\omega(A, E)$ such that

$$(x_i , A) = \lambda(u_i , A) - (1 - \lambda)(w_i , A)$$

Note that $V_\omega \cap C_\omega(A, E) = V_{\omega D} \cap C_\omega(A, E)$, and that

$$\prod(g), \prod(h) \in \prod(V_\omega^o \cap C_\omega(A, E)') \subset \prod(V_{\omega D}^o \cap C_\omega(A , E)') = V_{\omega D}^p \cap m_\infty^+(A, E').$$

We obtain

$$\begin{aligned}|<(x_i , A),\prod(f)>| &\leq \lambda|<(u_i , A),\prod(f)>| + (1 - \lambda)|<(w_i , A),\prod(f)>| \\ &\leq \max(<(u_i , A),\prod(g)>, <(w_i , A),\prod(h)>) \leq 1 ,\end{aligned}$$

then $\prod(f) \in V_{\omega D}^p$.

(vi) $V_{\omega D}^o \cap C_\omega(A, E)' = V_{\omega S}^o \cap C_\omega(A , E)'$.

This follows from (iv) and the fact that $V_{\omega D} \subset V_{\omega S}$.

(vii) $M_{\infty S} \subseteq V_{\omega S}^p \subseteq \pi(V_{\omega S}^o)$.

Clearly $V_{\omega S}^p \subseteq \pi(V_{\omega S}^o)$. Let $(x_i', A) \in M_{\infty,S}$ and let u' , in $V^o \cap C'$, be such that $-u' \leq x_i' \leq u'$ $(i \in A)$. Then $|<x, u'>| \leq q(x)$ $(x \in E)$. From this we can show, by a similar argument given in the proof of (2.3.1) and the definition of $q_{\omega S}$, that

$$|\sum_A <x_i , x_i'>| \leq q_{\omega S}(x_i , A) \quad \text{for all} \quad (x_i , A) \in \ell_\omega^1<A, E> .$$

(viii) $\prod(V_{\omega S}^o) \subseteq M_{\infty S}$.

Let f be in $V_{\omega S}^o$. Since $V_{\omega S}^o = (S(V_\omega))^o = S(V_\omega^o)$, there exists $g \in V_\omega^o \cap C_\omega(A, E)'$ such that $-g \leq f \leq g$, and thus

$$- \prod(g) \leq \prod(f) \leq \prod(g). \tag{3.12}$$

Note that $V_{\omega}^{o} \subset V_{\omega D}^{o}$. By (ii), we have $\prod(g) \in M_{\infty}^{+}$, hence there exists $u' \in V^{o} \cap C'$ such that $o \leqslant g \circ J_{i} \leqslant u'$ $(i \in A)$. We conclude from (3.12) that

$$-u' \leqslant -g \circ J_{i} \leqslant f \circ J_{i} \leqslant g \circ J_{i} \leqslant u' \quad \text{for all} \quad i \in A ,$$

thus $\prod(f) = (f \circ J_{i}, A) \in M_{\infty S}$.

(2.3.6) Corollary. Let $(E, C, \mathscr{P})$ be a locally solid space and let q be a continuous monotone seminorm on E . Suppose further that V is the unit ball of q and that

$$V_{\varepsilon D} = \{(x_{i}, A) \in \ell^{1}\langle A, E\rangle : q_{\varepsilon D}(x_{i}, A) < 1\}$$

$$V_{\varepsilon S} = \{(x_{i}, A) \in \ell^{1}\langle A, E\rangle : q_{\varepsilon S}(x_{i}, A) < 1\} .$$

Then we have

$$V_{\varepsilon D}^{o} \cap C_{\varepsilon}(A, E)' = V_{\varepsilon S}^{o} \cap C_{\varepsilon}(A, E)' = V_{\omega D}^{p} \cap m_{\infty}^{+}(A, E') \tag{3.13}$$

$$= \{(u_{i}', A) \in m_{\infty}^{+}(A, E') : \exists\, u' \in V^{o} \cap C' \text{ such that } o \leqslant u_{i}' \leqslant u' \quad \forall\, i \in A\} .$$

Proof. Denote by M_{∞}^{+} the set in the right-side of (3.13) . Since $q_{\varepsilon D}$ is the restriction of $q_{\omega D}$ on $\ell^{1}\langle A, E\rangle$, (3.3) holds for all $(x_{i}, A) \in \ell^{1}\langle A, E\rangle$, thus $M_{\infty}^{+} \subseteq V_{\varepsilon D}^{o} \cap C_{\varepsilon}(A, E)'$.

On the other hand, by the Hahn-Banach extension theorem (precisely (1.1.3)) any $g \in V_{\varepsilon D}^{o} \cap C_{\varepsilon}(A, E)'$ has an extension $f \in V_{\omega D}^{o} \cap C_{\omega}(A, E)'$ because of $q_{\varepsilon D} = q_{\omega D}$ on $\ell^{1}\langle A, E\rangle$. Denote by Q the quotient map from $(\ell_{\omega}^{1}\langle A, E\rangle)'$ onto $(\ell_{\omega}^{1}\langle A, E\rangle)'/(\ell^{1}\langle A, E\rangle)^{\perp}$. Since $\ker \prod = (\ell^{1}\langle A, E\rangle)^{\perp}$, where $\prod$ is the positive projection defined in (2.3.2), it follows that $g = Q(f)$, and hence from (2.3.5) (a) that

$$\widehat{\prod}(g) = \widehat{\prod}(Q(f)) = \widehat{\prod}(f) \in M_\infty^+ ,$$

where $\widehat{\prod}$ is the bijection associated with $\prod$. Therefore $V^o_{\varepsilon,D} \cap C_\varepsilon(A, E)' \subseteq M^+_\infty$.

Finally, the assertion that $V^o_{\varepsilon D} \cap C_\varepsilon(A, E)' = V^o_{\varepsilon S} \cap C_\varepsilon(A, E)' = V^p_{\omega D} \cap m^+_\infty(A, E')$ follows from (2.3.5)(a) and (c) by making use the fact that $m_\infty(A, E')$ is the topological dual of $(\ell^1\langle A, E\rangle, C_\varepsilon(A, E), \mathscr{P}_{\varepsilon D})$.

(2.3.7) Corollary. <u>Let</u> $(E, C, \mathscr{P})$ <u>be a locally convex Riesz space and</u> q <u>a continuous Riesz seminorm on</u> E . <u>Denote by</u> V <u>the unit ball of</u> q , <u>by</u> $\Sigma_{\omega S}$ <u>the closed unit ball of</u> $q_{\omega S}$ <u>in</u> $\ell^1_\omega\langle A, E\rangle$, <u>and by</u> $\Sigma_{\varepsilon S}$ <u>the closed unit ball of</u> $q_{\varepsilon S}$ <u>in</u> $\ell^1\langle A, E\rangle$. <u>Then</u> $q_{\omega S}$ <u>and</u> $q_{\varepsilon S}$ <u>are Riesz seminorms on</u> $\ell^1_\omega\langle A, E\rangle$ <u>and</u> $\ell^1\langle A, E\rangle$ <u>respectively, and</u>

$$\Sigma_{\omega S} = \{(x_i , A) \in \ell^1_\omega\langle A, E\rangle : q_\omega(\textstyle\sum_{i\in\alpha}|x_i|) \leq 1 \ \forall \alpha \in \mathfrak{F}(A)\} , \tag{3.13}$$

$$\Sigma_{\varepsilon S} = \{(x_i , A) \in \ell^1\langle A, E\rangle : q_\varepsilon(\textstyle\sum_{i\in\alpha}|x_i|) \leq 1 \ \forall \alpha \in \mathfrak{F}(A)\} , \tag{3.14}$$

$$\Sigma^p_{\omega S} = \Sigma^o_{\varepsilon S} = \{(x'_i , A) \in m_\infty(A, E') : \exists\, u' \in V^o \cap C' \text{ such that } |x'_i| \leq u' \ \ \forall\, i \in A\} . \tag{3.15}$$

<u>Proof</u>. As (E^A, C^A) is a Riesz space, $\ell^1_\omega\langle A, E\rangle$ and $\ell^1\langle A, E\rangle$ are solid subspaces of (E^A, C^A) , and as $C_\omega(A, E), C_\varepsilon(A, E)$ are the relative cones induced by C^A , it follows that $(\ell^1_\omega\langle A, E\rangle, C_\omega(A, E))$ and $(\ell^1\langle A, E\rangle , C_\varepsilon(A, E))$ are Riesz spaces. Clearly, $q_{\omega S}$ and $q_{\varepsilon S}$ are Riesz seminorms on $\ell^1_\omega\langle A, E\rangle$ and $\ell^1\langle A, E\rangle$ respectively, hence

$$q_{\omega S}(x_i , A) = q_{\omega S}(|x_i| , A) \quad \text{and} \quad q_{\varepsilon S}(x_i , A) = q_{\varepsilon S}(|x_i| , A) .$$

The equalities (3.13) and (3.14) then follow from (2.2.1). Since $\Sigma^o_{\varepsilon S}$ is solid, the equality (3.15) is a consequence of (2.3.6) and (2.3.5)(d) .

2.4 The topological dual of $m_o(A, E)$ and of $m_{o,2}(A, E)$

Let $(E, C, \mathcal{P})$ be a locally o-convex space, let $\mathcal{P}$ be determined by a family P of monotone seminorms and let C be generating. Then $E^{(A)} \subset m_\infty(A, E)$, $(E')^{(A)} \subset p_o(A, E')$ because C' is $\sigma(E', E)$-closed, and $\{q_{\infty B} : q \in P\}$ determines a locally solid topology $\mathcal{P}_\infty$ on $(m_\infty(A, E), m_\infty^+(A, E))$, where each $q_{\infty B}$ is given by

$$q_{\infty B}(x_\iota, A) = \inf\{\max(q(u), q(w)) : u, w \in C, -u \leq x_\iota \leq w \ \forall\ \iota \in A\} \quad (4.1)$$

(see (2.2.8)(2)). The $\mathcal{P}_\infty$-closure of $E^{(A)}$ in $m_\infty(A, E)$ is denoted by $m_o(A, E)$.

If $(u_\iota, A) \in m_\infty^+(A, E)$ and if $(u'_\iota, A) \in p_o^+(A, E')$, then there exists $u \in C$ and $u' \in C'$ such that

$$u_\iota \leq u \ (\iota \in A) \quad \text{and} \quad \sum_{\iota\in\alpha} u'_\iota \leq u' \quad \text{for all } \alpha \in \mathcal{F}(A),$$

hence we have

$$0 \leq \sum_{\iota\in\alpha} \langle u_\iota, u'_\iota\rangle \leq \sum_{\iota\in\alpha} \langle u, u'_\iota\rangle \leq \langle u, u'\rangle \quad \text{for all } \alpha \in \mathcal{F}(A);$$

consequently we obtain

$$\sum_A \langle u_\iota, u'_\iota\rangle = \sup\{\sum_{\iota\in\alpha} \langle u_\iota, u'_\iota\rangle : \alpha \in \mathcal{F}(A)\} < \infty .$$

Since $m_\infty^+(A, E) = m_\infty(A, E) \cap C^A$ and $p_o^+(A, E') = p_o(A, E') \cap (C')^A$ are generating cones in $m_\infty(A, E)$ and $p_o(A, E')$ respectively, it follows that $\langle m_\infty(A, E), p_o(A, E')\rangle$ is a dual pair under the bilinear form (3.1)(§2.3). Therefore $p_o(A, E')$ can be identified with a vector subspace of the algebraic dual of $m_\infty(A, E)$, and thus elements in $p_o^+(A, E')$ are positive linear functionals on $(m_\infty(A, E), m_\infty^+(A, E))$. Furthermore, elements in $p_o(A, E')$ are $\mathcal{P}_\infty$-continuous as the following result shows.

(2.4.1) Lemma. Let $(E, C, \mathcal{P})$ be a locally o-convex space and let C be generating. Then $p_0(A, E')$ may be identified with a subspace of the topological dual $m_\infty(A, E)'$ of $(m_\infty(A, E), m_\infty^+(A, E), \mathcal{P}_\infty)$, hence elements in $p_0^+(A, E')$ are positive $\mathcal{P}_\infty$-continuous linear functionals on $m_\infty(A, E)$.

Proof. It is sufficient to show that elements in $p_0(A, E')$ are $\mathcal{P}_\infty$-continuous. Let $(u_i', A) \in p_0(A, E')$ and let u', in C', be such that $\sum_{i\in\alpha} u_i' \leq u'$ for all $\alpha \in \mathcal{F}(A)$. Then there exists a continuous monotone seminorm q on E such that

$$|\langle x, u'\rangle| \leq q(x) \quad \text{for all} \quad x \in E. \tag{4.2}$$

The $\mathcal{P}_\infty$-continuity of (u_i', A) will follow by showing that

$$\left|\sum_A \langle x_i, u_i'\rangle\right| \leq q_{\infty B}(x_i, A) \quad \text{for all} \quad (x_i, A) \in m_\infty(A, E). \tag{4.3}$$

In fact, for any $(x_i, A) \in m_\infty(A, E)$ and $\delta > 0$, there exist, by (4.1), u and w in C with $-u \leq x_i \leq w$ $(i \in A)$ such that

$$\max(q(u), q(w)) < q_{\infty B}(x_i, A) + \delta .$$

For any $\alpha \in \mathcal{F}(A)$, we have

$$\sum_{i\in\alpha} \langle x_i, u_i'\rangle \leq \sum_{i\in\alpha} \langle w, u_i'\rangle \leq \langle w, u'\rangle \leq q(w) ,$$

$$\sum_{i\in\alpha} \langle -x_i, u_i'\rangle \leq \sum_{i\in\alpha} \langle u, u_i'\rangle \leq \langle u, u'\rangle \leq q(u) .$$

It then follows that

$$\left|\sum_{i\in\alpha} \langle x_i, u_i'\rangle\right| \leq \max(q(w), q(u)) < q_{\infty B}(x_i, A) + \delta .$$

As δ was arbitrary, there is

$$\left|\sum_{i\in\alpha} \langle x_i, u_i'\rangle\right| \leq q_{\infty B}(x_i, A) \quad \text{for all} \quad \alpha \in \mathcal{F}(A) ,$$

which obtains the required inequality (4.3).

(2.4.2) Lemma. Let $(E, C, \mathcal{P})$ be a locally solid space. There exists a positive projection $\prod_\infty$ from the topological dual $m_\infty(A, E)'$ of

$(m_\infty(A, E), m_\infty^+(A, E), \mathcal{P}_\infty)$ <u>onto</u> $p_0(A, E')$ <u>such that</u>

$$\ker \prod_\infty = (m_0(A, E))^\perp \quad \underline{\text{and}} \quad \prod_\infty(m_\infty^+(A, E)') = p_0^+(A, E') ,$$

<u>where</u> $(m_0(A, E))^\perp$ <u>is the polar of</u> $m_0(A, E)$ <u>taken in</u> $m_\infty(A, E)'$.

<u>Proof</u>. For each $\iota \in A$, as $(E, C, \mathcal{P})$ is locally solid, it follows from (2.2.8)(3) that the natural map J_ι is a positive continuous (injective) linear map from $(E, C, \mathcal{P})$ into $(m_\infty(A, E), m_\infty^+(A, E), \mathcal{P}_\infty)$, and hence that

$$x'_\iota = f \circ J_\iota \in E' \quad \text{for all} \quad f \in m_\infty(A, E)' .$$

Therefore we may define a map $\prod_\infty : m_\infty(A, E)' \to ((E')^A, (C')^A)$ by setting

$$\prod_\infty(f) = (f \circ J_\iota , A) \quad \text{for all} \quad f \in m_\infty(A, E)' . \qquad (4.4)$$

Clearly, π_∞ is linear and positive since each J_ι is positive.

In order to verify that $\prod_\infty(m_\infty(A, E)') \subseteq p_0(A, E')$, it is sufficient to show that $\prod_\infty(m_\infty^+(A, E)') \subseteq p_0^+(A, E')$ because $m_\infty^+(A, E)'$ is generating. Let $g \in m_\infty^+(A, E)'$ and let q be a continuous monotone seminorm on E such that

$$|\langle(x_\iota , A), g\rangle| \leq q_{\infty B}(x_\iota , A) \quad \text{for all} \quad (x_\iota , A) \in m_\infty(A, E) . \quad (4.5)$$

Then for any $x \in E$ and $\alpha \in \mathcal{F}(A)$, we have, by (2.2.8)(3), that

$$|\langle x, \textstyle\sum_{\iota\in\alpha} g \circ J_\iota\rangle| = |\langle J_\alpha(x), g\rangle| \leq q_{\infty B}(J_\alpha(x)) \leq q_B(x) ,$$

thus $\sum_{\iota\in\alpha} g \circ J_\iota \in (B(V))^o$, where q_B is the gauge of $B(V)$ which is a $\mathcal{P}$ -neighbourhood of 0 and V is the unit ball of q in E (see (1.2.5)). As $\sum_{\iota\in\alpha} g \circ J_\iota$ is directed upwards (with respect to $\alpha \in \mathcal{F}(A)$) , the $\sigma(E', E)$-compactness of $(B(V))^o$ insures that there exists $z' \in (B(V))^o$ such that $0 \leq \sum_{\iota\in\alpha} g \circ J_\iota \leq z'$ $(\alpha \in \mathcal{F}(A))$, and hence from $S(V) \subset B(V)$

that there exists $u' \in V^{o} \cap C'$ such that

$$\sum_{i\in\alpha} g \circ J_i \leqslant u' \quad \text{for all } \alpha \in \mathcal{F}(A) . \tag{4.6}$$

Therefore $\prod_{\infty}(g) = (g \circ J_i , A) \in p_o^+(A, E')$.

The proof can be completed by making use similar arguments given in the corresponding parts of the proof of (2.3.2).

(2.4.3) Theorem (Walsh [1]). Let $(E, C, \mathcal{P})$ be a locally solid space. Then $p_o(A, E')$ may be identified with the topological dual $m_o(A, E)'$ of $(m_o(A, E), m_o^+(A, E), \mathcal{P}_{\infty o})$.

Proof. Note that $\mathcal{P}_{\infty o}$ is the relative topology induced by $\mathcal{P}_{\infty}$; then we can complete the proof by a similar argument given in that of (2.3.3).

For a subset B of $m_{\infty}(A, E)$, we denote by B^p the polar of B taken in $p_o(A, E')$ (w.r.t the duality $<m_{\infty}(A, E), p_o(A, E')>$), and by B^o the polar of B taken in the topological dual $m_{\infty}(A, E)'$ of $(m_{\infty}(A, E), m_{\infty}^+(A, E), \mathcal{P}_{\infty})$. We have the following result which is analogous to (2.3.5).

(2.4.4) Theorem (Walsh [1]). Let $(E, C, \mathcal{P})$ be a locally solid space, let q be a continuous monotone seminorm on E , and let $\prod_{\infty}$ be the positive projection from $m_{\infty}(A, E)'$ onto $p_o(A, E')$ as defined in (2.4.2). Suppose further that V is the unit ball of q in E , and that

$$V_{\infty} = \{(x_i , A) \in m_{\infty}(A, E) : q_{\infty}(x_i , A) < 1\} ,$$
$$V_{\infty B} = \{(x_i , A) \in m_{\infty}(A, E) : q_{\infty B}(x_i , A) < 1\} ,$$
$$V_{\infty S} = \{(x_i , A) \in m_{\infty}(A, E) : q_{\infty S}(x_i , A) < 1\} .$$

Then the following statements hold:

(a) $\prod_\infty(V^o_{\infty B} \cap m^+_\infty(A, E)') = V^p_{\infty B} \cap p^+_o(A, E')$

$= \{(u'_i, A) \in p^+_o(A, E') : \exists\, u' \in V^o \cap C'$ such that $\sum_{i\in\alpha} u'_i \leqslant u'$
$\forall \alpha \in \mathcal{F}(A)\}$.

(b) $V^o_{\infty B} \cap m^+_\infty(A, E)' = V^o_{\infty S} \cap m^+_\infty(A, E)' = V^o_\infty \cap m^+_\infty(A, E)'$, consequently

$$V^p_{\infty B} \cap p^+_o(A, E') = V^p_{\infty S} \cap p^+_o(A, E') = V^p_\infty \cap p^+_o(A, E') .$$

(c) $\prod_\infty(V^o_{\infty B}) = V^p_{\infty B} = co\{-(V^p_{\infty B} \cap p^+_o(A, E')) \cup (V^p_{\infty B} \cap p^+_o(A, E'))\}$
$(=D(V^p_{\infty B}))$.

(d) $\prod_\infty(V^o_\infty) = V^p_\infty$ and

$\prod_\infty(V^o_{\infty S}) = V^p_{\infty S}$

$= \{(x'_i, A) \in p_o(A, E') : \exists\, u' \in V^o \cap C'$ such that
$-u' \leqslant \sum_{i\in\alpha} x'_i \leqslant u' \quad \forall \alpha \in \mathcal{F}(A)\}$.

Proof. For simplicity of notation, we write

$M^+_{o,B} = \{(u'_i, A) \in p^+_o(A, E') : \exists\, u' \in V^o \cap C'$ such that
$\sum_{i\in\alpha} u'_i \leqslant u' \quad \forall \alpha \in \mathcal{F}(A)\}$,

$M_{o,S} = \{(x'_i, A) \in p_o(A, E') : \exists\, u' \in V^o \cap C'$ such that
$-u' \leqslant \sum_{i\in\alpha} x'_i \leqslant u' \quad \forall \alpha \in \mathcal{F}(A)\}$.

As q_∞ is monotone and $q_\infty = q_{\infty D}$, V_∞ is postively order-convex and $V_\infty = D(V_\infty)$. Also (1.2.6) insures that $q_{\infty B}$ is the gauge of $B(V_\infty)$, $q_{\infty S}$ is the gauge of $S(V_\infty)$, and that

$$V_{\infty B} = B(V_\infty) \ , \ V_{\infty S} = S(V_\infty) \ ,$$
$$V_{\infty B} \cap m^+_\infty(A, E) = V_{\infty S} \cap m^+_\infty(A, E) = V_\infty \cap m^+_\infty(A, E) .$$

The proof is complete by considering the following steps:

(i) $M^+_{o,B} \subseteq V^p_{\infty B} \cap p^+_o(A, E')$.

Let $(u'_i, A) \in M^+_{o,B}$ and let u' , in $V^o \cap C'$, be such that $\sum_{i\in\alpha} u'_i \leq u' \ (\alpha \in \mathcal{F}(A))$. As $|\langle x, u'\rangle| \leq q(x) \ (x \in E)$, it follows from (4.3) that $(u'_i, A) \in V^p_{\infty B} \cap p^+_o(A, E')$.

(ii) $M^+_{o,B} = \prod_\infty (V^o_{\infty B} \cap m^+_\infty(A, E)')$.

Clearly $M^+_{o,B} \subseteq \prod_\infty (V^o_{\infty B} \cap m^+_\infty(A, E)')$. If $g \in V^o_{\infty B} \cap m^+_\infty(A, E)'$, then by the proof of (2.4.2), there exists $u' \in V^o \cap C'$ such that (4.6) holds, i.e.,

$$\sum_{i\in\alpha} g \circ J_i \leq u' \quad \text{for all} \quad \alpha \in \mathcal{F}(A) ,$$

hence $\prod_\infty(g) = (g \circ J_i , A) \in M^+_{o,B}$ as required.

(iii) $M^+_{o,B} = V^p_{\infty B} \cap p^+_o(A, E')$.

This follows from (i) and (ii) by making use of the fact that $\prod_\infty$ is a positive projection and $V^p_{\alpha B} \cap p^+_o(A, E') \subset V^o_{\infty B} \cap m^+_\infty(A, E)'$.

(iv) $V^o_\infty \cap m^+_\infty(A, E)' \subseteq V^o_{\infty B} \cap m^+_\infty(A, E)'$.

Since $V_\infty = D(V_\infty)$, it follows from Theorem (1.1.7) that $V^o_\infty = F(V^o_\infty)$, and hence from Proposition (1.1.11) that

$$V^o_{\infty B} = (B(V_\infty))^o = D(F(V^o_\infty)) = D(V^o_\infty) \tag{4.7}$$

Therefore we obtain

$$V^o_\infty \cap m^+_\infty(A, E) \subset D(V^o_\infty) \cap m^+_\infty(A, E)' = V^o_{\infty B} \cap m^+_\infty(A, E)' .$$

(v) The conclusion (b) holds.

Since $V_\infty = D(V_\infty) \subset V_{\infty S} \subset V_{\infty B}$, it follows from (iv) that

$$V^o_{\infty B} \cap m^+_\infty(A, E)' \subseteq V^o_{\infty S} \cap m^+_\infty(A, E)' \subseteq V^o_\infty \cap m^+_\infty(A, E)' \subseteq V^o_{\infty B} \cap m^+_\infty(A, E)' .$$

On the other hand, by (ii) and (iii), we have

$$V^p_{\infty B} \cap p^+_o(A, E') = \prod_\infty (V^o_{\infty B} \cap m^+_\infty(A, E)') = \prod_\infty (V^o_\infty \cap m^+_\infty(A, E)') \supseteq V^p_\infty \cap P^+_0(A, E')$$

it then follows that

$$V^p_\infty \cap p^+_0(A, E') \subseteq V^p_{\infty B} \cap p^+_o(A, E') \subseteq V^p_{\infty S} \cap p^+_o(A, E') \subseteq V^p_\infty \cap p^+_0(A, E') .$$

(vi) $\prod_\infty (V^o_{\infty B}) \subseteq D(V^p_{\infty B})$, and hence the conclusion (c) holds.

If $f \in V^o_{\infty B}$, by (4.7) there exists $\lambda \in (0, 1)$ and $g, h \in V^o_\infty \cap m^+_\infty(A, E)'$ such that $f = \lambda g - (1 - \lambda)h$, we conclude from (v), (ii) and (iii) that

$$\prod_\infty(f) = \lambda \prod_\infty(g) - (1 - \lambda)\prod_\infty(h) \in D(V^p_{\infty,B}) ;$$

therefore $\prod_\infty (V^o_{\infty B}) \subseteq D(V^p_{\infty B})$. Now it is easily seen that the conclusion (c) holds because of

$$D(V^p_{\infty B}) \subseteq V^p_{\infty B} \subseteq \prod_\infty (V^o_{\infty B}) \subseteq D(V^p_{\infty B}) .$$

(vii) $\prod_\infty (V^o_\infty) = V^p_\infty$.

It is sufficient to show that $\prod_\infty (V^o_\infty) \subseteq V^p_\infty$. Note that V^o_∞ is positively dominated. For any $f \in V^o_\infty$, there exist $g, h \in V^o_\infty \cap m^+_\infty(A, E)'$ such that $-h \leq f \leq g$, hence

$$-\prod_\infty(h) \leq \prod_\infty(f) \leq \prod_\infty(g)$$

and

$$\prod_\infty(g), \prod_\infty(h) \in \prod_\infty(V_\infty^o \cap m_\infty^+(A, E)') = V_\infty^p \cap p_0^+(A, E')$$

By a similar argument given in the proof of (v) of (2.3.5), we have $\prod_\infty(f) \in V_\infty^p$.

(viii) $\prod_\infty(V_{\infty S}^o) = V_{\infty S}^p = M_{o,S}$.

The proof is similar to that of corresponding parts in (2.3.5) .

Remark. According to the conclusion (b) and (4.7), we see that $V_{\infty B}^o$ is decomposable in the topological dual $m_\infty(A, E)'$ of $(m_\infty(A, E), m_\infty^+(A, E), \mathcal{P}_\infty)'$. Also the conclusion (b) insures that

$$V_{\infty S}^p \cap p_0^+(A, E') = \prod_\infty(V_{\infty S}^o \cap m_\infty^+(A, E)') = \prod_\infty(V_\infty^o \cap m_\infty^+(A, E)') = V_\infty^p \cap p_0^+(A, E').$$

(2.4.5) Corollary. Let $(E, C, \mathcal{P})$ be a locally solid space and q a continuous monotone seminorm on E . Suppose further that V is the unit ball of q in E , and that

$$V_{oB} = \{(x_i, A) \in m_o(A, E) : q_{\infty B}(x_i, A) < 1\} ,$$
$$V_{oS} = \{(x_i, A) \in m_o(A, E) : q_{\infty S}(x_i, A) < 1\} ,$$
$$V_o = \{(x_i, A) \in m_o(A, E) : q_\infty(x_i, A) < 1\} .$$

Then the following statements hold:

(a) $V_{oB}^o \cap m_o^+(A, E)' = V_{oS}^o \cap m_o^+(A, E)' = V_o^o \cap m_o^+(A, E)' = V_{\infty B}^p \cap C_o(A, E')$

$$= \{(u_i', A) \in C_o(A, E') : \exists\, u' \in V^o \cap C' \text{ such that } \sum_{i\in\alpha} u_i' \leq u' \ \ \forall \alpha \in \mathcal{F}(A)\} .$$

(b) $V_{oB}^o = D(V_{oB}^o)$.

Proof. This follows from the Hahn-Banach extension theorem and (2.4.4).

(2.4.6) Corollary. Let $(E, C, \mathcal{P})$ be a locally convex Riesz space and q a continuous monotone seminorm on E. Denote by V the unit ball of q in E, and by $\Sigma_{\infty S}$ the closed unit ball of $q_{\infty S}$ in $m_\infty(A, E)$. Then $q_{\infty S}$ is a Riesz seminorm on $m_\infty(A, E)$, and

$$\Sigma_{\infty S} = \{(x_\iota, A) \in m_\infty(A, E) : \inf\{q(u) : |x_\iota| \leq u \ \forall \iota \in A\} \leq 1\} \quad (4.8)$$

$$\Sigma^p_{\infty S} = \{(x'_\iota, A) \in p_0(A, E') : \exists\, u' \in V^o \cap C' \text{ such that}$$

$$\sum_{\iota \in \alpha} |x'_\iota| \leq u' \ \forall \alpha \in \mathcal{F}(A)\} . \quad (4.9)$$

Proof. Clearly $q_{\infty S}$ is a Riesz seminorm on the Riesz space $(m_\infty(A, E), m^+_\infty(A, E))$, and the formula (4.8) is a consequence of (2.2.8)(2). To prove the formula (4.9), we first note that the set in the right side of (4.9) is obviously contained in $\Sigma^p_{\infty S}$. On the other hand, if $(x'_\iota, A) \in \Sigma^p_{\infty S}$, then $(|x'_\iota|, A) \in p_0(A, E')$ and $(|x'_\iota|, A) \in \Sigma^o_{\infty S}$ because $\Sigma^o_{\infty S}$ is solid, hence by (2.4.4)(d), there exists $u' \in V^o \cap C'$ such that

$$\sum_{\iota \in \alpha} |x'_\iota| \leq u' \quad \text{for all} \quad \alpha \in \mathcal{F}(A) .$$

There are exact analogues of everything we have established for the duality $<m_\infty(A, E), p_o(A, E')>$ that are valid for the duality $<m_{\infty,2}(A, E), p_o(A, E')>$. We state the results but omit the proofs, which are almost the same as the proofs of (2.4.1) through (2.4.5).

(2.4.7) Lemma. Let $(E, C, \mathcal{P})$ be a locally o-convex space and let C be generating. Then $p_0(A, E')$ may be identified with the topological dual $m_{\infty,2}(A, E)'$ of $(m_{\infty,2}(A, E), m^+_{\infty,2}(A, E), \mathcal{P}_{\infty,2})$, hence $p^+_0(A, E') \subset m^+_{\infty,2}(A, E)'$.

Denote by $m_{0,2}(A, E)'$ the topological dual of $(m_{0,2}(A, E), m^+_{0,2}(A, E), \mathcal{P}_{\infty,o,2})$, and by $(m_{0,2}(A, E))^{\perp}$ the polar of $m_{0,2}(A, E)$ taken in $m_{\infty,2}(A, E)'$. Then we have the following result which is anologous to (2.4.2) and (2.4.3).

(2.4.8) Theorem. Let $(E, C, \mathcal{P})$ be a locally solid space. There exists a positive projection $\prod_{\infty,2}$ from $m_{\infty,2}(A, E)'$ onto $p_0(A, E')$ such that

$$\ker \prod_{\infty,2} = (m_{0,2}(A, E))^{\perp} \quad \text{and} \quad \prod_{\infty,2}(m^+_{\infty,2}(A, E)') = p^+_0(A, E') .$$

Furthermore, if in addition, C is $\mathcal{T}$-closed, then $p_0(A, E')$ may be identified with $m_{0,2}(A, E)'$.

For a subset B of $m_{\infty,2}(A, E)$, we denote by B^p the polar of B taken in $p_0(A, E')$ (w.r.t. $<m_{\infty,2}(A, E), p_o(A, E')>$), and by B^o the polar of B taken in $m_{\infty,2}(A, E)'$.

(2.4.9) Theorem. Let $(E, C, \mathcal{P})$ be a locally solid space, let q be a continuous monotone seminorm on E, and let $\prod_{\infty,2}$ be the positive projection from $m_{\infty,2}(A, E)'$ onto $p_0(A, E')$ as defined in (2.4.8). Suppose further that V is the unit ball of q in E, and that

$$V_{\infty,2} = \{(x_i, A) \in m_{\infty,2}(A, E) : q''_{\infty}(x_i, A) < 1\} ,$$

$$V_{\infty,2,B} = \{(x_i, A) \in m_{\infty,2}(A, E) : q''_{\infty\,B}(x_i, A) < 1\} ,$$

$$V_{\infty,2,S} = \{(x_i, A) \in m_{\infty,2}(A, E) : q''_{\infty\,S}(x_i, A) < 1\} .$$

Then the following statements hold:

(a) $\prod_{\infty,2}(V^{o}_{\infty,2,B} \cap m^{+}_{\infty,2}(A, E)') = V^{p}_{\infty,2,B} \cap p^{+}_{o}(A, E')$

$= \{(u'_{i}, A) \in p^{+}_{o}(A, E') : \exists\, u' \in V^{o} \cap C'$ such that

$\sum_{i \in \alpha} u'_{i} \leq u' \quad \forall\, \alpha \in \mathcal{F}(A)\}$.

(b) $V^{o}_{\infty,2,B} \cap m^{+}_{\infty,2}(A, E)' = V^{o}_{\infty,2,S} \cap m^{+}_{\infty,2}(A, E)' = V^{o}_{\infty,2} \cap m^{+}_{\infty,2}(A, E)'$,

consequently $V^{p}_{\infty,2,B} \cap p^{+}_{o}(A, E') = V^{p}_{\infty,2,S} \cap p^{+}_{o}(A, E') =$

$V^{p}_{\infty,2} \cap p^{+}_{o}(A, E')$.

(c) $\prod_{\infty,2}(V^{o}_{\infty,2,B}) = V^{p}_{\infty,2,B} = D(V^{p}_{\infty,2,B})(\text{in } (p_{o}(A, E'), p^{+}_{o}(A, E')))$.

(2.4.10) Corollary. Let $(E, C, \mathcal{P})$ be a locally solid space, let C be closed and let q be a continuous monotone seminorm on E . Suppose further that V is the unit ball of q in E , and that

$$V_{o,2} = V_{\infty,2} \cap m_{o,2}(A, E)\ , \quad V_{o,2,B} = V_{\infty,2,B} \cap m_{o,2}(A, E)\ ,$$

$$V_{o,2,S} = V_{\infty,2,S} \cap m_{o,2}(A, E)\ .$$

Then the following statements hold:

(a) $V^{o}_{o,2,B} \cap m^{+}_{o,2}(A, E)' = V^{o}_{o,2,S} \cap m^{+}_{o,2}(A, E)' = V^{o}_{o,2} \cap m^{+}_{o,2}(A, E)'$

$= V^{p}_{\infty,2,B} \cap p^{+}_{o}(A, E')$

$= \{(u'_{i}, A) \in p^{+}_{o}(A, E') : \exists\, u' \in V^{o} \cap C'$ such that

$\sum_{i \in \alpha} u'_{i} \leq u' \quad \forall\, \alpha \in \mathcal{F}(A)\}$.

(b) $V^{o}_{o,2,B} = D(V^{o}_{o,2,B})$.

Let $(E, C, \mathcal{P})$ be a locally solid space and let C be closed. It is known that the canonical embedding map of $m_{o}(A, E)$ into $m_{o,2}(A, E)$ is continuous. In view of (2.4.3) and (2.4.8), $m_o(A, E)' = m_{o,2}(A, E)' = p_o(A, E')$. On the other hand, if q is a continuous monotone seminorm on E, then we obtain, by (2.4.5) and (2.4.10), that

$$V^o_{oB} = V^o_{o,2,B} .$$

Therefore $q''_{\infty B} = q_{\infty B}$ on $m_o(A, E)$. Furthermore, we have:

(2.4.11) Proposition. <u>Let</u> $(E, C, \mathcal{P})$ <u>be a locally solid space</u>, <u>let</u> C <u>be closed and let</u> q <u>be a continuous monotone seminorm on</u> E. <u>Then</u>

$$q''_{\infty B} = q_{\infty B} \quad \text{on } m_o(A, E) .$$

<u>Consequently</u>, <u>the canonical embedding map of</u> $m_o(A, E)$ <u>into</u> $m_{o,2}(A, E)$ <u>is a topological isomorphism from the first space into the second</u>, <u>while</u> $E^{(A)}_{\infty}$ <u>and</u> $E^{(A)}_{\infty,2}$ <u>are topological isomorphic</u> (<u>under the identify map</u>). <u>If</u> $(m_o(A, E), \mathcal{P}_{\infty o})$ <u>is complete</u> (<u>in particular</u>, $(E, \mathcal{P})$ <u>is a Fréchet space</u>), <u>then the canonical embedding map is surjective</u>.

<u>Proof</u>. The first assertion has been observed in the above. If $(E, \mathcal{P})$ is a Fréchet space, then $(m_\infty(A, E), m^+_\infty(A, E), \mathcal{P}_\infty)$ is complete, and $m_o(A, E)$ is $\mathcal{P}_\infty$-closed in $m_\infty(A, E)$ because $m_o(A, E)$ is the $\mathcal{P}_\infty$-closure of $E^{(A)}$ in $m_\infty(A, E)$; consequently, $(m_o(A, E), \mathcal{P}_{\infty o})$ is complete. Suppose now that $(m_o(A, E), \mathcal{P}_{\infty o})$ is complete. Then $m_o(A, E)$ is $\mathcal{P}_{\infty,2}$-closed in $m_{\infty,2}(A, E)$ by the first assertion, and hence $m_o(A, E) = m_{o,2}(A, E)$ because $m_{o,2}(A, E)$ is the $\mathcal{P}_{\infty,2}$-closure of $m_o(A, E)$ in $m_{\infty,2}(A, E)$.

CHAPTER 3. SOME CHARACTERIZATIONS OF THE TOPOLOGY OF UNIFORM CONVERGENCE ON ORDER-BOUNDED SETS

For a locally solid space $(E, C, \mathcal{P})$, it is known from Proposition (1.3.15) that the locally solid topology $\sigma_S(E, E')$ associated with $\sigma(E, E')$ is the topology of uniform convergence on all order-intervals. In particular, if $(E, C, \mathcal{P})$ is a locally convex Riesz space, then $\sigma_S(E, E')$ is called the Diendonné topology (see Wong and Ng [1]), and it is shown (see Wong and Ng [1]) that $(E, C, \sigma_S(E, E'))$ is complete if and only if $(E, C, \sigma_S(E, E'))$ is boundedly and locally order complete. Therefore it is interesting to seek some characterizations for the topology $\mathcal{P}$ to be $\sigma_S(E, E')$. The purpose of this chapter is to give such characterizations in terms of some special continuous linear maps, among them the most important one is perhaps cone-absolutely summing maps.

The concept of cone-absolutely summing maps was first considered by Schaefer and his school in the Banach lattices case (see Schaefer [3;2]), and has been extended by Walsh [1] and others to the case of locally o-convex spaces with closed and generating cones. Walsh [1] was very successful in extending Schlotterbeck's results to a fairly general setting, and offering counter example to show that some of his results really involve the lattice structure intrinsically.

3.1 Cone-absolutely summing mappings

Let $(X, \mathcal{P})$ and $(Y, \mathcal{Z})$ be locally convex spaces and A a non-empty index set. Denote by $L^*(X, Y)$ the space consisting of all linear maps from

X into Y, and by $L(X(\mathcal{P}), Y(\mathcal{I}))$ the space consisting of all <u>continuous</u> linear maps from $(X, \mathcal{P})$ into $(Y, \mathcal{I})$. If the topologies on X and Y do not require special notations, we write $L(X, Y)$ for $L(X(\mathcal{P}), Y(\mathcal{I}))$.

If $T \in L^*(X, Y)$, then we can associate a linear map $T^A : X^A \to Y^A$ defined by

$$T^A(x_i, A) = (Tx_i, A) \quad \text{for all} \quad (x_i, A) \in X^A .$$

Denote by $T^{(A)}$ the restriction of T^A to $X^{(A)}$. Then $T^{(A)}(X^{(A)}) \subset Y^{(A)}$, and thus $T^{(A)} \in L^*(X^{(A)}, Y^{(A)})$. For each $\iota \in A$, $T^{(A)} \circ J_\iota$ is a linear map from X into $J_\iota(Y)$ with

$$\pi_j((T^{(A)} \circ J_\iota)x) = \begin{cases} Tx & \text{if } j = \iota \\ 0 & \text{if } j \neq \iota , \end{cases}$$

where J_ι is the natural map, and π_j is the j-th projection; it then follows that

$$\pi_\iota \circ T^{(A)} \circ J_\iota = T . \tag{1.1}$$

On the other hand, if $T \in L(X, Y)$, then T^A is always a continuous map from $(X^A, \mathcal{P}^A)$ into $(Y^A, \mathcal{I}^A)$.

Let $(E, C, \mathcal{P})$ be an ordered convex space and Y a locally convex space. An $T \in L^*(E, Y)$ is said to be <u>cone-absolutely summing</u> if $T^{(\mathbb{N})}$ is a continuous map from $C^{(\mathbb{N})}$, equipped with the relative topology induced by $\mathcal{P}_\varepsilon$, into $Y_\pi^{(\mathbb{N})}$; clearly T is cone-absolutely summing if and only if for any continuous seminorm q on Y there exists a continuous seminorm p on E such that the inequality

$$\sum_{i=1}^n q(Tu_i) \leq p\left(\sum_{i=1}^n u_i\right) \tag{1.2}$$

holds for any finite subset $\{u_1, u_2, \ldots, u_n\}$ of C. It then follows that T is cone-absolutely summing if and only if for any non-empty index set A, $T^{(A)}$ is a continuous map from $C^{(A)}$, equipped with the relative topology induced by $\mathcal{P}_\varepsilon$, into $Y_\pi^{(A)}$. The set consisting of all cone-absolutely summing mappings from E into Y, denoted by $L^\ell(E, Y)$, is obviously a vector subspace of $L^*(E, Y)$. On the other hand, if $T \in L^\ell(E, Y)$, then Formula (1.2) implies that T is a continuous map from C into Y; in particular, if $(E, C, \mathcal{P})$ is locally decomposable, then T must be continuous, and thus $L^\ell(E, Y) \subset L(E, Y)$.

Moreover, if $(E, C, \mathcal{P})$ is a locally solid space, then $E_{\varepsilon,D}^{(\mathbb{N})}$ is a locally solid space by Lemma (2.2.13), and $\mathcal{P}_{\varepsilon D}$ coincides with $\mathcal{P}_\varepsilon$ on $C^{(\mathbb{N})}$; consequently, T is cone-absolutely summing if and only if $T^{(\mathbb{N})} \in L(E_{\varepsilon,D}^{(\mathbb{N})}, Y_\pi^{(\mathbb{N})})$, and this is the case if and only if $T^{(A)} \in L(E_{\varepsilon,D}^{(A)}, Y_\pi^{(A)})$ for any non-empty index set A.

(3.1.1) Proposition. <u>Let</u> $(E, C, \mathcal{P})$ <u>be a locally solid space</u>, $(Y, \mathfrak{T})$ <u>a locally convex space and suppose that</u> $T \in L(E, Y)$. <u>Then the following statements are equivalent</u>:

(a) $T \in L^\ell(E, Y)$.

(b) $T^A \in L(\ell^1\langle A, E\rangle, \ell^1[A, Y])$ <u>for any non-empty index set</u> A.

(c) <u>The adjoint map</u> T' <u>of</u> T <u>sends equicontinuous subsets of</u> Y' <u>into order-intervals in</u> E'.

(d) <u>For any continuous seminorm</u> q <u>on</u> Y, <u>there exists</u> $f \in C'$ <u>such that</u>

$$q(Tx) \leq \sup\{g(x) : g \in [-f, f]\}$$
$$= \inf\{f(w) : w \pm x \in C\} \qquad \text{for any} \quad x \in E.$$

(e) $T \in L(E(\sigma_S), Y)$.

<u>Proof</u>. Clearly (d) and (e) are equivalent in view of Proposition (1.3.15)(b).

(a) ⇒ (b) : Formula (1.2) implies that T^A is a continuous map from $C_\varepsilon(A, E)$, equipped with the relative topology induced by $\mathcal{P}_\varepsilon$, into $(\ell^1[A, Y], \mathcal{T}_\pi)$. As $\mathcal{P}_{\varepsilon D}$ coincides with $\mathcal{P}_\varepsilon$ on $C_\varepsilon(A, E)$ and $\mathcal{P}_{\varepsilon D}$ is locally decomposable, it follows that $T^A \in L(\ell^1\langle A, E\rangle, \ell^1[A, Y])$.

(b) ⇒ (c) : Let M be an equicontinuous subset of Y'. By Theorem (2.1.3), the family $\{y' : y' \in M\}$ is a continuous linear functional on $(\ell^1[M, Y], \mathcal{T}_\pi)$, hence $(T^M)'(\{y' : y' \in M\}) = \{T'y' : y' \in M\}$ is a continuous linear functional on $(\ell^1\langle M, E\rangle, C_\varepsilon(M, E), \mathcal{P}_{\varepsilon D})$. As $m_\infty(M, E')$ is identified with $(\ell^1\langle M, E\rangle, C_\varepsilon(M, E), \mathcal{P}_{\varepsilon D})'$ (see Theorem (2.3.3)), it follows from the definition of the space $m_\infty(M, E')$ that $T'(M)$ is an order-interval in E'.

(c) ⇒ (d) : Let q be a continuous seminorm on Y and $U = \{y \in Y : q(y) \leq 1\}$. Then there exists an $f \in C'$ such that $T'(U^\circ) \subset [-f, f]$, hence

$$q(Tx) = \sup\{\langle Tx, y'\rangle : y' \in U^\circ\} \leq \sup\{g(x) : g \in [-f, f]\} \quad (x \in E) .$$

In view of Proposition (1.3.15)(b), we have

$$\sup\{g(x) : g \in [-f, f]\} = \inf\{f(u) : u \pm x \in C\}$$

which proves the implication.

(d) ⇒ (a) : Let q be a continuous seminorm on Y and $U = \{y \in Y : q(y) \leq 1\}$. By the assumption, there exists an $f \in C'$ such that
that

$$q(Tx) \leq \inf\{f(w) : w \pm x \in C\}. \tag{1.3}$$

Putting

$$p(x) = \inf\{f(w) : w \pm x \in C\} \quad \text{for any} \quad x \in E .$$

Then p is a $\sigma_S(E, E')$-continuous seminorm on E such that

$$p(w) = f(w) \quad \text{for all} \quad w \in C , \tag{1.4}$$

hence p is $\mathcal{P}$-continuous. On the other hand, for any finite subset $\{u_1, \ldots, u_n\}$ of C, we have from (1.3) and (1.4) that

$$\Sigma_{i=1}^{n} q(Tu_i) \leq \Sigma_{i=1}^{n} p(u_i) = f(\Sigma_{i=1}^{n} u_i) = p(\Sigma_{i=1}^{n} u_i)$$

which shows that $T \in L^{\ell}(E, Y)$.

It is remarkable that the vector space $L^{\ell}(E, Y)$ does not depend upon the locally solid topology on E, but only on the dual pair $\langle E, E'\rangle$ (since (a) is equivalent with (e)), and that $T \in L^{\ell}(E, Y)$ if and only if $T^{\mathbb{N}} \in L(\ell^1\langle E\rangle, \ell^1[Y])$.

The equivalence of (b) and (c) in the preceding result is due to Walsh [1] . If $(E, C, \mathcal{P})$ is a locally convex Riesz space and if $f \in C'$, then

$$\langle |x|, f\rangle = \sup\{g(x) : g \in [-f, f]\} \quad \text{for any} \quad x \in E$$

because E' is a solid subspace of the order-bounded dual E^b of (E, C). Therefore we obtain the following

(3.1.2) Corollary. <u>Let</u> $(E, C, \mathcal{P})$ <u>be a locally convex Riesz space</u> <u>and</u> Y <u>a locally convex space</u>. <u>Then an</u> $T \in L(E, Y)$ <u>is cone-absolutely summing if and only if for any continuous seminorm</u> q <u>on</u> Y <u>there exists a</u>

$f \in C'$ <u>such that</u>

$$q(Tx) \leq \langle |x|, f\rangle \quad \underline{\text{for all}} \quad x \in E .$$

Let $(E, C, \mathcal{P})$ be a locally solid space and $(Y, \mathcal{T})$ a locally convex space. Suppose that $T \in L(E, Y)$ and $T^{\mathbb{N}} \in L^*(\ell^1\langle E\rangle, \ell^1[Y])$. Then $T^{\mathbb{N}}$ is a continuous linear map from $(E^{\mathbb{N}}, C^{\mathbb{N}}, \mathcal{P}^{\mathbb{N}})$ into $(Y^{\mathbb{N}}, \mathcal{T}^{\mathbb{N}})$, and the graph of $T^{\mathbb{N}}$ is closed for the relative topology on $\ell^1\langle E\rangle$ induced by $\mathcal{P}^{\mathbb{N}}$ and $\mathcal{T}_\pi$ because the relative topology on $\ell^1[Y]$ induced by $\mathcal{T}^{\mathbb{N}}$ is coarser than $\mathcal{T}_\pi$, hence the graph of $T^{\mathbb{N}}$ is closed for the product topology $\mathcal{P}_{\varepsilon D} \times \mathcal{T}_\pi$ since the relative topology on $\ell^1\langle E\rangle$ induced by $\mathcal{P}^{\mathbb{N}}$ is coarser than $\mathcal{P}_\varepsilon$. Moreover, if $(\ell^1\langle E\rangle, \mathcal{P}_{\varepsilon D})$ is barrelled and if $(\ell^1[Y], \mathcal{T}_\pi)$ is an infra-Pták space, then Robertson-Robertson's closed graph theorem ensures that $T^{\mathbb{N}} \in L(\ell^1\langle E\rangle, \ell^1[Y])$. In particular, if $(E, \mathcal{P})$ and $(Y, \mathcal{T})$ and Fréchet spaces, we obtain:

(3.1.3) Corollary. <u>Let</u> $(E, C, \mathcal{P})$ <u>be a Fréchet locally solid space</u>, $(Y, \mathcal{T})$ <u>a Fréchet space and</u> $T \in L(E, Y)$. <u>Then</u> $T \in L^{\ell}(E, Y)$ <u>if and only if</u> $T^{\mathbb{N}} \in L^*(\ell^1\langle E\rangle, \ell^1[Y])$, <u>and if and only if</u> T <u>sends every positive summable sequence in</u> E <u>into an absolutely summable sequence in</u> Y (<u>i.e</u>., $T^{\mathbb{N}}(C_\varepsilon(E)) \subset \ell^1[Y])$.

<u>Proof</u>. Under the hypotheses, $(\ell^1\langle E\rangle, \mathcal{P}_{\varepsilon D})$ and $(\ell^1[Y], \mathcal{T}_\pi)$ are Fréchet spaces and surely barrelled and infra-Pták spaces. The last part is obvious because $C_\varepsilon(E)$ is generating and $T^{\mathbb{N}}$ is linear.

The following result should be compared with Scháefer [3, (iv.2.7) p.242] .

(3.1.4) Corollary. <u>Let</u> $(E, C, \mathcal{P})$ <u>be a locally solid space and</u>

I <u>the identity map on</u> E . <u>Then the following statements are equivalent.</u>

(a) $I \in L^{\ell}(E, E)$.

(b) <u>Every equicontinuous subset of</u> E' <u>is order-bounded</u>.

(c) <u>For any continuous seminorm</u> p <u>on</u> E <u>there exists an</u> $f \in C'$ <u>such that</u>

$$p(x) \leq \sup\{g(x) : g \in [-f, f]\} = \inf\{f(w) : w \pm x \in C\} \quad (x \in E) .$$

(d) $\mathcal{P} = \sigma_S(E, E')$.

(e) <u>The canonical embedding map</u> φ <u>of</u> $(\ell^1_a\langle E\rangle, C_\pi(E), \mathcal{P}_{\pi D})$ <u>into</u> $(\ell^1\langle E\rangle, C_\varepsilon(E), \mathcal{P}_{\varepsilon D})$ <u>is a topological isomorphism from the first space onto the second one</u>.

<u>Furthermore, if in addition,</u> $(E, C, \mathcal{P})$ <u>is metrizable</u> (<u>resp</u>. <u>Fréchet</u>) <u>then</u> $I \in L^{\ell}(E, E)$ <u>if and only the canonical embedding map of</u> $(\ell^1[E], C_\pi(E), \mathcal{P}_\pi)$ <u>into</u> $(\ell^1\langle E\rangle, C_\varepsilon(E), \mathcal{P}_{\varepsilon D})$ <u>is a topological isomorphism</u> (<u>resp</u>. <u>algebraic isomorphism</u>) <u>from the first space onto the second</u>.

<u>Proof</u>. The equivalence between (a), (b), (c) and (d) is a consequence of Proposition (3.1.1).

(a) $\Rightarrow$ (e) : For any continuous Riesz seminorm q on E , there exists a continuous Riesz seminorm p on E such that the inequality

$$\Sigma_{i=1}^{n} q(u_i) \leq p(\Sigma_{i=1}^{n} u_i)$$

holds for any finite subset $\{u_1, \ldots, u_n\}$ of C , hence

$$C_\varepsilon(E) = C_\pi(E) \quad \text{and} \quad q_\pi \leq p_\varepsilon \quad \text{on} \quad C_\varepsilon(E) .$$

As I is positive, it then follows from the definitions of $q_{\pi D}$ and $p_{\varepsilon D}$ that $I^{\mathbb{N}}$ is a continuous bijective linear map from $(\ell^1\langle E\rangle, C_\varepsilon(E), \mathcal{P}_{\varepsilon D})$ onto $(\ell^1_a\langle E\rangle, C_\pi(E), \mathcal{P}_{\pi D})$. Clearly the inverse map $(I^{\mathbb{N}})^{-1}$ of $I^{\mathbb{N}}$ is φ , hence φ is bijective and φ^{-1} is continuous. As φ is always continuous, the implication follows.

(e) ⇒ (a) : If φ is a topological isomorphism from $(\ell^1_a\langle E\rangle, C_\pi(E), \mathcal{P}_{\pi D})$ onto $(\ell^1\langle E\rangle, C_\varepsilon(E), \mathcal{P}_{\varepsilon D})$, then the inverse φ^{-1} of φ is $I^{\mathbb{N}}$. As the relative topology on $\ell^1_a\langle E\rangle$ induced by $\mathcal{P}_\pi$ is coarser than $\mathcal{P}_{\pi D}$, it follows that $I^{\mathbb{N}} \in L(\ell^1\langle E\rangle, \ell^1[E])$, and hence that $I \in L^\ell(E, E)$.

Finally, if $(E, C, \mathcal{P})$ is metrizable, then

$$\ell^1[E] = \ell^1_a\langle E\rangle \quad \text{and} \quad \mathcal{P}_\pi = \mathcal{P}_{\pi D} ,$$

by Lemma (2.2.2); if $(E, C, \mathcal{P})$ is a Fréchet space and if φ is an algebraic isomorphism from $\ell^1_a\langle E\rangle$ onto $\ell^1\langle E\rangle$, then $I^{\mathbb{N}} = \varphi^{-1}$ and hence $I^{\mathbb{N}}$ is a linear map from $\ell^1\langle E\rangle$ onto $\ell^1_a\langle E\rangle = \ell^1[E]$. However the conclusions follow from this corollary or Corollary (3.1.3) .

(3.1.5) Lemma. <u>Let</u> $(E, C, \mathcal{P})$ <u>and</u> $(F, K, \mathcal{L})$ <u>be ordered convex spaces</u>, <u>let</u> X, Y <u>be locally convex spaces</u>. <u>Then the following statements hold</u>:

(1) <u>If</u> $T \in L^\ell(E, X)$ <u>and</u> $S \in L(X, Y)$, <u>then</u> $S \circ T \in L^\ell(E, Y)$.

(2) <u>If</u> $S \in L(E, F)$ <u>is positive and</u> $T \in L^\ell(F, Y)$, <u>then</u> $T \circ S \in L^\ell(E, Y)$.

<u>Proof</u>. (1) Let r be a continuous seminorm on Y and let q be a continuous seminorm on X such that

$$r(Sy) \leq q(y) \quad \text{for all} \quad y \in X .$$

Since $T \in L^{\ell}(E, X)$, there exists a continuous seminorm p on E such that the inequality

$$\Sigma_{i=1}^{n} q(Tu_i) \leq p(\Sigma_{i=1}^{n} u_i)$$

holds for any finite subset $\{u_1, \ldots, u_n\}$ of C. We conclude from

$$\Sigma_{i=1}^{n} r(S(Tu_i) \leq \Sigma_{i=1}^{n} q(Tu_i) \leq p(\Sigma_{i=1}^{n} u_i)$$

that $S \circ T \in L^{\ell}(E, Y)$.

(2) Let r be a continuous seminorm on Y and let q be a continuous seminorm on F such that the inequality

$$\Sigma_{i=1}^{n} r(Tu_i) \leq q(\Sigma_{i=1}^{n} y_i)$$

holds for any finite subset $\{y_1, \ldots, y_n\}$ of K.

Let p be a continuous seminorm on E such that

$$q(Sx) \leq p(x) \quad \text{for all} \quad x \in E .$$

If $\{u_1, u_2, \ldots, u_n\}$ is any finite subset of C, then the positivity of S insures that $\{Su_1, \ldots, Su_n\}$ is a finite subset of K, and thus

$$\Sigma_{1=1}^{n} r(T(Su_i)) \leq q(\Sigma_{i=1}^{n} Su_i) \leq p(\Sigma_{i=1}^{n} u_i) .$$

This shows that $T \circ S \in L^{\ell}(E, Y)$.

Throughout this chapter X and Y always denote locally convex spaces, E and F always denote locally solid spaces, while G and H will denote locally convex Riesz spaces, unless a statement is made to the contrary.

Let G and H be locally convex Riesz space and let H be order complete. Then the space consisting of all order-bounded (regular in the terminology of Schaefer [3]) linear maps from G into H, denoted by $L^b(G, H)$, is an order complete Riesz space under the canonical ordering, but $L(G, H)$ need not be a Riesz space. If H is both locally and boundedly order complete, then $L^\ell(G, H)$ is an ℓ-ideal (i.e., solid subspace) in $L^b(G, H)$ as the following result shows:

(3.1.6) Theorem. <u>Let</u> $(G, C, \mathcal{P})$ <u>and</u> $(H, K, \mathcal{Q})$ <u>be locally convex Riesz spaces. If</u> $(H, K, \mathcal{Q})$ <u>is both locally and boundedly order complete, then</u> $L^\ell(G, H)$ <u>is an</u> ℓ<u>-ideal in</u> $L^b(G, H)$, <u>and hence is an order complete Riesz space in its own right</u>.

<u>Proof</u>. Let $T \in L^\ell(G, H)$. For any continuous Riesz seminorm q on H, there exists a continuous Riesz seminorm p on G such that the inequality

$$\Sigma_{i=1}^{n} q(Tw_i) \leq p(\Sigma_{i=1}^{n} w_i)$$

holds for any finite subset $\{w_1, \ldots, w_n\}$ of C. Let $u \in C$ and let $u = \Sigma_{i=1}^{n} u_i$ with $u_i \in C$. As q is a Riesz seminorm, there is

$$q(\Sigma_{i=1}^{n} |Tu_i|) \leq \Sigma_{i=1}^{n} q(Tu_i) \leq p(u) . \tag{1.5}$$

Hence for any fixed $u \in C$, the set $\{\Sigma_{i=1}^{n} |Tu_i| : u_i \in C, \Sigma_{i=1}^{n} u_i = u\}$ is bounded in H, and directed upwards by the Riesz decomposition property. Therefore the bounded order completeness of H implies that the supremum

$$|T|(u) = \sup\{\Sigma_{i=1}^{n} |Tu_i| : u_i \in C, \Sigma_{i=1}^{n} u_i = u\}$$

exists for each $u \in C$, thus $|T|$ exists in $L^b(G, H)$, consequently $L^\ell(G, H) \subset L^b(G, H)$.

The next step is going to verify that $|T| \in L^\ell(G, H)$. Let r be a continuous Riesz seminorm on H. As $(H, K, \gtrsim)$ is locally order complete, there exists a o-neighbourhood W in H which is order complete, and a closed absolutely convex, solid o-neighbourhood V in H such that

$$V \subseteq W \subseteq \{y \in F : r(y) \leq 1\} .$$

Now if q is the gauge of V, then q is a continuous Riesz seminorm on H such that the inequality

$$r(\sup B) \leq \sup\{q(b) : b \in B\}$$

holds for any bounded subset B of H which is directed upwards. In particular,

$$r(|T|u) \leq \sup\{q(\Sigma_{i=1}^{n} |Tu_i|) : u_i \in C , \Sigma_{i=1}^{n} u_i = u\}$$

holds for any $u \in C$. For the seminorm q, there exists a continuous seminorm p on E such that Formula (1.5) holds for any finite subset $\{w_1, \ldots, w_n\}$ of C since $T \in L^\ell(G, H)$. It then follows that the inequalities

$$\begin{aligned}
\Sigma_{j=1}^{n} r(|T|w_j) &\leq \Sigma_{j=1}^{n} \sup\{q(\Sigma_{i=1}^{m_j} |Tu_{ij}|) : u_{ij} \in C , \Sigma_{i=1}^{m_j} u_{ij} = w_j\} \\
&= \sup\{\Sigma_{j=1}^{n} q(\Sigma_{i=1}^{m_j} |Tu_{ij}|) : u_{ij} \in C , \Sigma_{i=1}^{m_j} u_{ij} = w_j\} \\
&\leq \sup\{\Sigma_{j=1}^{n} \Sigma_{i=1}^{m_j} q(Tu_{ij}) : u_{ij} \in C , \Sigma_{i=1}^{m_j} u_{ij} = w_j\} \\
&\leq \sup\{p(\Sigma_{j=1}^{n} \Sigma_{i=1}^{m_j} u_{ij}) : u_{ij} \in C , \Sigma_{i=1}^{m_j} u_{ij} = w_j\} \\
&= p(\Sigma_{j=1}^{n} w_j)
\end{aligned}$$

hold for any finite subset $\{w_1, \ldots, w_n\}$ of C . Thus $|T| \in L^{\ell}(G, H)$.

From the above two assertions we conclude that $L^{\ell}(G, H)$ is a Riesz space under the canonical ordering. Finally, it is easy to verify that $L^{\ell}(G, H)$ is a solid subset of $L^b(G, H)$. Therefore $L^{\ell}(G, H)$ is an ℓ-ideal in $L^b(G, H)$.

(3.1.7) Corollary. <u>For locally convex Riesz spaces</u> $(G, C, \mathcal{P})$ <u>and</u> $(H, K, \mathcal{I})$, <u>the following assertions hold</u>:

(a) <u>If</u> $\mathcal{P} = \sigma_S(G, G')$ <u>and if</u> $(H, K, \mathcal{I})$ <u>is both locally and boundedly order complete</u>, <u>then</u> $L(G, H)$ <u>is an</u> ℓ<u>-ideal in</u> $L^b(G, H)$.

(b) <u>If</u> $\mathcal{I} = \sigma_S(H, H')$ <u>and if</u> H <u>is</u> $\mathcal{I}$<u>-complete, then</u> $L^{\ell}(G, H)$ <u>is an</u> ℓ<u>-ideal in</u> $L^b(G, H)$.

<u>Proof</u>. The conclusion (a) follows from Lemma (3.1.5), Theorem (3.1.6) and Corollary (3.1.4); while the conclusion (b) is a consequence of Theorem (3.1.5) because $(H, K, \mathcal{I})$ is, under the hypotheses, locally and boundedly order complete in view of Wong and Ng [1, Theorem (13.9)].

In particular, if $(G, C, \|\cdot\|)$ is a normed vector lattice such that the norm $\|\cdot\|$ is additive on C , then the norm-topology coincides with $\sigma_S(G, G')$. Therefore the conclusion (a) of the preceding result is an improvement of Krengel's result (see Peressini [1, (IV. 3.8) p.174].

It is known from Wong and Ng [1, Theorems (13.9) and (13.13)] that $(H, K, \sigma_S(H, H'))$ is complete if and only if $H \cong (H')^b_n$, where $(H')^b_n$ is the space consisting of all normal integrals on H' . Therefore, if

$(H')^b_n \subseteq H$, in view of the preceding corollary, we obtain the following result, essential due to Peressini [2] .

(3.1.8) Corollary. <u>Let</u> $(G, C, \mathcal{P})$ <u>and</u> $(H, K, \mathcal{I})$ <u>be locally convex Riesz spaces</u>. <u>If</u> $(H')^b_n \subseteq H$, <u>then</u> $L(G(\sigma_S), H(\sigma_S))$ <u>is an</u> ℓ<u>-ideal in</u> $L^b(G, H)$.

Let $L^o(G'(\sigma_S), H)$ be the space consisting of all linear maps $T : G' \to H$ that send some $\sigma_S(G', G)$-neighbourhood of 0 in G' into an order-bounded subset of H . Then $L^o(G'(\sigma_S), H) \subset L(G'(\sigma_S), H)$. The following result shows that $G \otimes H$ can be identified with a vector subspace of $L^o(G'(\sigma_S), H)$.

(3.1.9) Lemma. <u>Let</u> $(G, C, \mathcal{P})$ <u>and</u> $(H, K, \mathcal{I})$ <u>be locally convex Riesz spaces and let</u> $\mathcal{P}'$ <u>be a locally solid topology on</u> G' <u>which is consistent with the dual pair</u> $\langle G', G''\rangle$. <u>Then</u>

$$G \otimes H \subset L^o(G'(\sigma_S), H) \subset L^\ell(G'(\mathcal{P}'), H) ,$$

<u>and dually</u>

$$G' \otimes H \subset L^o(G(\sigma_S), H) \subset L^\ell(G, H) .$$

<u>Proof</u>. Each element $v = \Sigma_{i=1}^n x_i \otimes y_i$ in $G \otimes H$ defines a continuous linear map $T \in L(G'(\sigma_S), H)$ by virtue of

$$f \rightsquigarrow T(f) = \Sigma_{i=1}^n \langle x_i , f\rangle\, y_i \quad \text{for all} \quad f \in G' .$$

Here we may assume that n is the rank of v . Let

$$u = \sup\{|x_i| : 1 \leq i \leq n\} \text{ and } w = \sup\{|y_i| : 1 \leq i \leq n\} .$$

Then $T([-u, u]^o) \subset n^2[-w, w]$; in fact, if $f \in [-u, u]^o$, we have, for any $g \in [-w, w]^o$, that

$$|\langle T(f), g\rangle| \leq \Sigma_{i=1}^{n} |\langle x_i , f\rangle\langle y_i , g\rangle| \leq n^2 ,$$

which obtains the required conclusion in view of the bipolar theorem. Therefore $T \in L^o(G'(\sigma_S), H)$.

On the other hand, since $\sigma_S(G', G) \leq \sigma_S(G', G'')$ and $\sigma_S(G', G'')$ is the coarsest locally solid topology on G' consistent with $\langle G', G''\rangle$, it follows from Proposition (3.1.1) that

$$L^o(G'(\sigma_S), H) \subseteq L^o(G'(\sigma_S(G', G'')), H) \subseteq L(G'(\sigma_S(G', G''), H) = L^{\ell}(G'(\mathcal{P}'), H).$$

This proves the first part. The proof of the second part is similar, therefore the proof is complete.

If $(H, K, \mathfrak{T})$ is order complete, Peressini [2, Proposition 3] has shown that $L^o(G'(\sigma_S), H)$ is an ℓ-ideal in $L^b(G', H)$. Furthermore, $L^o(G'(\sigma_S), H)$ is the ℓ-ideal in $L^b(G', H)$ generated by $G \otimes H$ as the following result shows.

(3.1.10) Proposition. <u>Let</u> $(G, C, \mathcal{P})$ <u>and</u> $(H, K, \mathfrak{T})$ <u>be locally convex Riesz spaces</u>, <u>let</u> (H, K) <u>be order complete and let</u> K <u>be</u> $\mathfrak{T}$<u>-closed</u>. <u>Then</u> $L^o(G'(\sigma_S), H)$ <u>is the</u> ℓ-<u>ideal in</u> $L^b(G', H)$ <u>generated by</u> $G \otimes H$. <u>Dually</u>, $L^o(G(\sigma_S), H)$ <u>is the</u> ℓ-<u>ideal in</u> $L^b(G, H)$ <u>generated by</u> $G' \otimes H$.

<u>Proof</u>. In view of Lemma (3.1.9) and the above remark, it has only to show that for any $T \in L^o(G'(\sigma_S), H)$ there exists $u \in C$ and $w \in K$ such that

$$-(u \otimes w) \leq T \leq u \otimes w ;$$

but this is equivalent to verify that

$$f(u)w \pm T(f) \in K \quad \text{for all} \quad f \in C' .$$

In fact, since $T \in L^o(G'(\sigma_S), H)$, there exists $u \in C$ and $w \in K$ such that $T([-u, u]^o) \subset [-w, w]$. For any fixed $f \in C'$ and any $\varepsilon > 0$, we have that $f/_{(f(u)+\varepsilon)}$ is in $[-u, u]^o$, hence

$$(f(u) + \varepsilon)w \pm T(f) \in K$$

because of $f(u) + \varepsilon > 0$. As K is closed and ε is arbitrary, it is true that $f(u)w \pm T(f) \in K$, which obtains our assertion.

As $G \otimes H$ is always identified with the vector subspace of $L(G'(\sigma_S), H)$ consisting of elements of finite rank, the preceding result can be stated as the following form: If $(H, K, \mathcal{Q})$ is order complete and K is $\mathcal{Q}$-closed, then $L^o(G'(\sigma_S), H)$ is the ℓ-ideal in $L^b(G', H)$ generated by the vector subspace of $L(G'(\sigma_S), H)$ of finite rank, and dually, $L^o(G(\sigma_S), H)$ is the ℓ-ideal in $L^b(G, H)$ generated by the subspace of $L(G, H)$ of finite rank.

Let $(G, C, \mathcal{P})$ be a locally decomposable space, let $(H, K, \mathcal{Q})$ be a locally o-convex space and let $L^+(G, H)$ be the cone consisting of all continuous positive linear maps of G into H , i.e., $L^+(G, H) = \{T \in L(G, H) : T(C) \subset K\}$. Setting

$$L^r(G, H) = L^+(G, H) - L^+(G, H) .$$

Then $L^r(G, H)$ is always an absolutely dominated subspace of $L(G, H)$ as well as of $L^*(G, H)$. On the other hand, the local decomposability of $(G, C, \mathcal{P})$ insures that $L^r(G, H)$ is an order-convex subspace of $L(G, H)$ as well as of $L^*(G, H)$. Thus $L^r(G, H)$ is a solid subspace of $L(G, H)$.

Moreover, if $(G, C, \mathscr{P})$ and $(H, K, \mathfrak{I})$ are locally convex Riesz spaces, and if (H, K) is order complete, then

$$L^r(G, H) = \{T \in L^b(G, H) : |T| \in L(G, H)\} ,$$

hence $L^r(G, H)$ is an ℓ-ideal in $L^b(G, H)$, consequently $(L^r(G, H), L^+(G, H))$ is an order complete Riesz space.

From now on we always assume that $(G, C, \mathscr{P})$ and $(H, K, \mathfrak{I})$ are locally convex Riesz spaces, and that (H, K) is order complete. Let $\mathcal{V}$ be a neighbourhood base at 0 for $\mathfrak{I}$ consisting of closed, absolutely convex, solid sets in H , and let $\mathfrak{B}$ be a family consisting of closed, absolutely convex, solid bounded subsets of G such that $\cup \mathfrak{B} = G$. For each $B \in \mathfrak{B}$ and $V \in \mathcal{V}$, let

$$M_r(B, V) = \{T \in L^r(G, H) : |T|(B) \subseteq V\} .$$

Then $M_r(B, V)$ is an absolutely convex, <u>solid</u>, absorbing subsets of $L^r(G, H)$, hence $\{M_r(B, V) : B \in \mathfrak{B} , V \in \mathcal{V}\}$ determines a unique <u>locally solid</u> topology on $L^r(G, H)$; consequently $L^r(G, H)$ equipped with this topology, denoted by $L^r_{\mathfrak{B}_r}(G, H)$, is an order complete locally convex Riesz space. If $q^{(r)}_{(B,V)}$ denotes the gauge of $M_r(B, V)$, then it is easily seen that

$$q^{(r)}_{(B,V)}(T) = \sup\{|\langle |T|(x), f\rangle| : x \in B, f \in V^o\} \quad (T \in L^r(G, H)).$$

If $\mathfrak{B}$ is the family of all bounded subsets of G , then we write $L^r_{\beta_r}(G, H)$ for $L^r_{\mathfrak{B}_r}(G, H)$.

Similarly, for any $B \in \mathfrak{B}$ and $V \in \mathcal{V}$, let

$$M(B, V) = \{T \in L^r(G, H) : T(B) \subseteq V\} .$$

Then $M(B, V)$ is an absolutely convex, positive order-convex, absorbing subset of $L^r(G, H)$, hence $\{M(B, V) : B \in \mathcal{B}, V \in \mathcal{V}\}$ determines a locally o-convex topology on $L^r(G, H)$, consequently $L^r_{\mathcal{B}}(G, H)$ equipped with this topology is an order complete locally o-convex Riesz space. If $q_{(B,V)}$ denotes the gauge of $M(B, V)$, then

$$q_{(B,V)}(T) = \sup\{|\langle Tx, f\rangle| : x \in B, f \in V^o\} \quad (T \in L^r(G, H)) ,$$

and hence

$$q_{(B,V)}(|T|) = q^{(r)}_{(B,V)}(T) \quad \text{for all} \quad T \in L^r(G, H)$$

If $\mathcal{B}$ is the family of all bounded subsets of G, then we write $L^r_\beta(G, H)$ for $L^r_{\mathcal{B}}(G, H)$.

As $M_r(B, V) \subset M(B, V)$, it follows that the identity map is continuous from $L^r_{\mathcal{B}_r}(G, H)$ onto $L^r_{\mathcal{B}}(G, H)$.

Suppose now that $(H, K, \mathfrak{T})$ is both locally and boundedly order complete. Then Theorem (3.1.6) implies that $L^\ell(G, H)$ is an ℓ-ideal in $L^r(G, H)$. Denoting by $L^\ell_{\mathcal{B}_r}(G, H)$ the subspace of $L^r_{\mathcal{B}_r}(G, H)$, and by $L^\ell_{\mathcal{B}}(G, H)$ the subspace of $L^r_{\mathcal{B}}(G, H)$. Then $L^\ell_{\mathcal{B}_r}(G, H)$ is an order complete locally convex Riesz space, $L^\ell_{\mathcal{B}}(G, H)$ is an order complete locally o-convex Riesz space, and the identity map : $L^\ell_{\mathcal{B}_r}(G, H) \to L^\ell_{\mathcal{B}}(G, H)$ is continuous. If $\mathcal{P} = \sigma_S(G, G')$, then proposition (3.1.1) implies that

$$L^\ell(G, H) = L^r(G, H) = L(G, H) .$$

We now summarize our results as follows:

(3.1.11) Lemma. Let $(G, C, \mathcal{P})$ and $(H, K, \mathfrak{T})$ be locally convex

Riesz spaces, let $(H, K, \mathcal{I})$ be both locally and boundedly order complete. Then $L^{\ell}_{\beta_r}(G, H)$ is an order complete locally convex Riesz space. If in addition, $\mathcal{P} = \sigma_S(G, G')$, then

$$L^{\ell}(G, H) = L^{r}(G, H) = L(G, H) ,$$

and $L_{\beta_r}(G, H)$ is an order complete locally convex Riesz space.

Using a similar argument, we are able to define absolutely summing maps on a locally convex space. Let X and Y be locally convex spaces. An $T \in L^*(X, Y)$ is said to be absolutely summing if $T^{(\mathbb{N})} \in L(X_{\varepsilon}^{(\mathbb{N})}, Y_{\pi}^{(\mathbb{N})})$. Clearly T is absolutely summing if and only if for any continuous seminorm q on Y , there exists a continuous seminorm p on X such that the inequality

$$\Sigma_{i=1}^{n} q(Tx_i) \leqslant \sup\{\Sigma_{i=1}^{n} |\langle x_i , x'\rangle| : x' \in V_p^{o}\} \qquad (1.6)$$

holds for any finite subset $\{x_1, \ldots, x_n\}$ of X , where $V_p = \{x \in X : p(x) \leqslant 1\}$. In view of the remark after Lemma(2.1.6), T is absolutely summing if and only if for any continuous seminorm q on Y , there exists a continuous seminorm p on X such that the inequality

$$\Sigma_{i=1}^{n} q(Tx_i) \leqslant \sup\{p(\Sigma_{i=1}^{n} c_i x_i) : |c_i| = 1\} \qquad (1.7)$$

holds for any finite subset $\{x_1, \ldots, x_n\}$ of X . It then follows that T is absolutely summing if and only if $T^{(A)} \in L(X_{\varepsilon}^{(A)}, Y_{\pi}^{(A)})$ for any non-empty index set A . The set consisting of all absolutely summing maps from X into Y , denoted by $L^{s}(X, Y)$, is obviously a vector subspace of $L(X, Y)$.

Analogues to Proposition (3.1.1) holds for absolutely summing maps.

(3.1.12) Proposition. Let X and Y be locally convex spaces and $T \in L(X, Y)$. Then the following statements are equivalent.

(a) $T \in L^s(X, Y)$.

(b) $T^A \in L(\ell^1(A, X), \ell^1[A, Y])$ for any non-empty index set A .

(c) The adjoint map T' of T sends equicontinuous subsets of Y' into prenuclear subsets of X' .

(d) For any continuous seminorm q on Y there exists a $\sigma(X', X)$-closed equicontinuous subset B of X' and a positive Radon measure μ on B such that

$$q(Tx) \leq \int_B |\langle x, x'\rangle| d\mu(x') \quad \text{for all} \quad x \in X . \tag{1.8}$$

Proof. (a) ⇒ (b) : For any continuous seminorm q on Y there exists a continuous seminorm p on X such that the inequality

$$\Sigma_{i=1}^n q(Tz_i) \leq \sup\{\Sigma_{i=1}^n |\langle z_i, x'\rangle| : x' \in V_p^o\}$$

holds for any finite subset $\{z_1, \ldots, z_n\}$ of X , where $V_p = \{x \in X : p(x) \leq 1\}$. For any non-empty index set A and any $(x_i, A) \in \ell^1(A, X)$, we have, for any $\alpha \in \mathfrak{F}(A)$, that

$$\Sigma_{i\in\alpha} q(Tx_i) \leq \sup\{\Sigma_{i\in\alpha} |\langle x_i, x'\rangle| : x' \in V_p^o\} ;$$

it then follows that

$$q_\pi(Tx_i, A) = \sup_{\alpha\in\mathfrak{F}(A)} \Sigma_{i\in\alpha} q(Tx_i) \leq \sup_{\alpha\in\mathfrak{F}(A)} \sup\{\Sigma_{i\in\alpha} |\langle x_i, x'\rangle| : x' \in V_p^o\}$$

$$\leq \sup\{ \sup_{\alpha\in\mathfrak{F}(A)} \Sigma_{i\in\alpha} |\langle x_i, x'\rangle| : x' \in V_p^o\} = p_\varepsilon(x_i, A) ,$$

thus $T^A(x_i, A) = (Tx_i, A) \in \ell^1[A, Y]$ and $T^A \in L(\ell^1(A, X), \ell^1[A, Y])$.

(b) ⇒ (c) : Let M be an equicontinuous subset of Y' . By Theorem (2.1.3), the family $\{y' : y' \in M\}$ is a continuous linear functional on $\ell^1[M, Y]$, hence $(T^M)'(\{y' : y' \in M\}) = \{T'y' : y' \in M\}$ is a continuous linear functional on $\ell^1(A, X)$, and thus $T'(M)$ is a prenuclear subset of X' by Theorem (2.1.4).

(c) ⇒ (d) : Let q be a continuous seminorm on Y and $U = \{y \in Y : q(y) \leq 1\}$. Then $T'(U^o)$ is a prenuclear subset of X' , hence there exists a $\sigma(X', X)$-closed equicontinuous subset B of X' and a positive Radon measure μ on B such that

$$|\langle x, T'y'\rangle| \leq \int_B |\langle x, x'\rangle| \, d\mu(x') \quad \text{for all} \quad x \in X \text{ and } y' \in U^o .$$

It then follows that

$$q(Tx) \leq \int_B |\langle x, x'\rangle| \, d\mu(x') \quad \text{for all} \quad x \in X .$$

(d) ⇒ (a) : For any continuous seminorm q on Y , there exists a $\sigma(X', X)$-closed equicontinuous subset B of X' and a positive Radon measure μ on B such that Formula (1.8) holds. Let $V = B^o$. Then V is a convex o-neighbourhood in X such that V^o is the $\sigma(X', X)$-closed convex hull of $B \cup \{0\}$. For any finite subset $\{x_1, \ldots, x_n\}$ of X , we have

$$\begin{aligned} \Sigma_{i=1}^n q(Tx_i) &\leq \int_B \Sigma_{i=1}^n |\langle x_i, x'\rangle| \, d\mu(x') \\ &\leq \mu(B) \sup\{\Sigma_{i=1}^n |\langle x_i, x'\rangle| : x' \in V^o\} . \end{aligned}$$

Therefore $T \in L^s(X, Y)$.

It is worthwhile to notice that $T \in L^s(X, Y)$ if and only if $T^{\mathbb{N}} \in L(\ell^1(X), \ell^1[Y])$.

(3.1.13) Corollary. *If* X *is metrizable and if* $T \in L(X, Y)$, *then* $T \in L^s(X, Y)$ *if and only if* $T^{\mathbb{N}} \in L^*(\ell^1(X), \ell^1[Y])$, *i.e.*, T *sends every summable sequence in* X *into an absolutely summable sequence in* Y.

Proof. The necessity is obvious. Conversely, in view of Pietsch [1, (2.1.3)] $T^{\mathbb{N}} \in L(\ell^1(X), \ell^1[Y])$, and hence $T \in L^s(X, Y)$ be Proposition (3.1.12).

(3.1.14) Corollary. *Let* X *be a locally convex space and let* I *be an identity map on* X. *Then the following statements are equivalent.*

(a) $I \in L^s(X, X)$.

(b) *Every equivontinuous subset of* X' *is a prenuclear subset of* X'.

(c) *For any continuous seminorm* r *on* X, *there exists a* $\sigma(X', X)$-*closed equicontinuous subset* B *of* X' *and a positive Radon measure* μ *on* B *such that*

$$r(x) \leq \int_B |\langle x, x'\rangle| \, d\mu(x') \quad \text{for all} \quad x \in X .$$

(d) *The canonical embedding map* φ *of* $\ell^1[X]$ *into* $\ell^1(X)$ *is a topological isomorphism from the first space onto the second*.

Moreover, *if in addition*, X *is metrizable*, *then* $I \in L^s(X, X)$ *if and only if every summable sequence in* X *is absolutely summable*.

Proof. The equivalence between (a), (b) and (c) is a consequence of Proposition (3.1.12).

(a) $\Rightarrow$ (d) : In view of Proposition (3.1.12), $I^{\mathbb{N}} \in L(\ell^1(X), \ell^1[X])$.

As $\ell^1[X] \subset \ell^1(X)$, it follows that $I^{\mathbb{N}}$ is bijective. Clearly the inverse map $(I^{\mathbb{N}})^{-1}$ of $I^{\mathbb{N}}$ is φ, hence φ is bijective and φ^{-1} is continuous. As φ is always continuous, the implication follows.

(d) $\Rightarrow$ (a) : According to the assumption, the inverse map φ^{-1} of φ is $I^{\mathbb{N}}$, hence $I^{\mathbb{N}} \in L(\ell^1(X), \ell^1[X])$ which shows that $I \in L^s(X, X)$.

The final conclusion follows from Corollary (3.1.13) and this corollary.

Analogues to Lemma (3.1.5) holds for absolutely summing maps. The proof of the following result is routine, and hence will be omitted.

(3.1.15) Lemma. Let X, Y and Z be locally convex spaces. Suppose that $T \in L(X, Y)$ and $S \in L(Y, Z)$. If one of T and S is absolutely summing, then $S \circ T \in L^s(X, Z)$.

In view of the definitions of cone-absolutely summing maps and absolutely summing maps, we have:

(3.1.16) Lemma. Let $(E, C, \mathcal{P})$ be an ordered convex space and Y a locally convex space. Then $L^s(E, Y) \subset L^{\ell}(E, Y)$.

We shall give an example (see Examples (3.2.16)(b)) to show that $L^s(E, Y)$ is not necessarily equal to $L^{\ell}(E, Y)$. But, when E has some additional hypotheses, then they are equal as the following result shows.

(3.1.17) Proposition. Let $(E, C, \mathcal{P})$ be a Fréchet locally solid space for which C is closed, and let $C_{\varepsilon}(E)$ be generating. Then $L^{\ell}(E, Y) = L^s(E, Y)$.

Proof. In view of Corollary (2.2.5), $(\ell^1(E), C_\varepsilon(E), \mathcal{P}_\varepsilon)$ is a locally solid space, and hence

$$\ell^1(E) = \ell^1\langle E\rangle \quad \text{and} \quad \mathcal{P}_\varepsilon = \mathcal{P}_{\varepsilon D} \,. \tag{1.9}$$

For any $T \in L^{\ell}(E, Y)$, then $T^{\mathbb{N}} \in L(\ell^1\langle E\rangle, \ell^1[Y])$ by Proposition (3.1.1), hence $T^{\mathbb{N}} \in L(\ell^1(E), \ell^1[Y])$ in view of Formula (1.9), consequently, $T \in L^s(E, Y)$ by Proposition (3.1.12). This shows that $L^{\ell}(E, Y) \subseteq L^s(E, Y)$. Therefore the result follows from Lemma (3.1.16).

3.2 Some special classes of seminorms

Let $(E, C, \mathcal{P})$ be an ordered convex space for which C is generating, and let q be a seminorm on E. Let us say temporarily that q is an (PL)-seminorm if there exists an $f \in C'$ such that

$$q(x) \leq \inf\{f(w) : w \pm x \in C\} \quad (x \in E) \,. \tag{2.1}$$

If $(E, C, \mathcal{P})$ is a locally solid space, then Proposition (1.3.15) implies that

$$\inf\{f(w) : w \pm x \in C\} = \sup\{g(x) : g \in [-f, f]\} \quad (x \in E) \,,$$

hence q is an (PL)-seminorm if and only if there exists an $f \in C'$ such that

$$q(x) \leq \sup\{g(x) : g \in [-f, f]\} \quad (x \in E) \,.$$

It is also clear that every (PL)-seminorm on a locally solid space $(E, C, \mathcal{P})$ must be continuous because it is $\sigma_S(E, E')$-continuous. If $f \in C'$, then the seminorm p_f defined by

$$p_f(x) = |f(x)| \quad \text{for all} \quad x \in E \,,$$

is an (PL)-seminorm on E.

In order to give some characterizations of (PL)-seminorms in terms of the notion of cone-absolutely summing mappings, we require the following terminology: Let V be an absolutely convex, absorbing subset of E and q_V the gauge of V. Denote by $N(V)$ the kernel of q_V, by Q_V the quotient map from E onto $E/_{N(V)}$, and by $x(V)$ (or $\hat{x}$) the equivalent class $x + N(V)$ modulo $N(V)$. Then the quotient seminorm of q_V, denoted by $\hat{q}_V$, is actually a norm on $E/_{N(V)}$, hence $(E/_{N(V)}, \hat{q}_V)$ is a normed space. Moreover, we have

$$\hat{q}_V(x(V)) = \inf\{q_V(x + z) : z \in N(V)\} = q_V(x) ;$$

$$\{x(V) \in E/_{N(V)} : \hat{q}_V(x(V)) < 1\} \subset Q_V(V) \subset \{x(V) \in E/_{N(V)} : \hat{q}_V(x(V)) \leq 1\} .$$

Furthermore, if V is positively order-convex, then $N(V)$ is an order-convex subspace of E, hence $(E/_{N(V)}, Q_V(C), \hat{q}_V)$ is an ordered normed space. For simplicity of notation, we denote by E_V the normed space or the ordered normed space just introduced. If q is a seminorm on E and if $V = \{x \in E : q(x) \leq 1\}$, we let $Q_q = Q_V$ and $E_q = E_V$. If V and W are absolutely convex absorbing subsets of E such that $V \subseteq W$, then the canonical map from E_V onto E_W is denoted by $Q_{W,V}$. Hence we have the relation

$$Q_W = Q_{W,V} \circ Q_V .$$

Dually, if Y is a locally convex space and if B is an absolutely convex bounded subset of Y, then $\bigcup_{n \geq 1} nB$ is the vector subspace of Y generated by B, and the gauge p_B of B defined on $\bigcup_{n\geq 1} nB$ is a norm, hence we denote by $Y(B)$ the normed space $(\bigcup_{n\geq 1} nB, p_B)$. Clearly the relative topology on $Y(B)$ is coarser than the p_B-topology, it follows that the (naturally) imbedding map, denoted by j_B, is a continuous linear map from $Y(B)$ into Y.

(3.2.1) Lemma. Let $(E, C, \mathcal{P})$ be a locally solid space, let q be a seminorm on E and let $V = \{x \in E : q(x) \leq 1\}$. Then the following statements are equivalent.

(a) q is an (PL)-seminorm.

(b) The quotient map $Q_q : E \to E_q$ is cone-absolutely summing.

(c) V^o is an order-bounded subset of E' .

(d) There exists a continuous monotone seminorm r on E with $q \leq r$ such that the canonical map $Q_{q,r} : E_r \to E_q$ is cone-absolutely summing.

Proof. As

$$q(x) = \hat{q}(Q_q(x)) \quad \text{and} \quad V^o = Q_q'((Q_q(V))^o) ,$$

it follows from Proposition (3.1.1) that (a), (b) and (c) are equivalent. In view of Lemma (3.1.5), (d) implies (b). Therefore we complete the proof by showing that (a) implies (d). Let $f \in C'$ be such that

$$q(x) \leq \inf\{f(w) : w \pm x \in C\} \quad (x \in E)$$

and define

$$r(x) = \inf\{f(w) : w \pm x \in C\} \quad (x \in E) .$$

Then r is a continuous monotone seminorm on E such that

$$q \leq r \quad \text{and} \quad r(u) = f(u) \quad \text{for all} \quad u \in C ,$$

hence E_r is an ordered normed space. If $\{Q_r(u_1), \ldots, Q_r(u_n)\}$ is any finite subset of $Q_r(C)$, we may assume without loss of generality that $u_j \in C$ (in fact, if $Q_r(y) \in Q_r(C)$, then there exists an $w \in C$ such that $Q_r(y) = Q_r(w)$, hence $r(y) = r(u))$, thus we have

$$\Sigma_{i=1}^n \hat{q}(Q_{q,r}(Q_r(u_i)) = \Sigma_{i=1}^n q(u_i) \leq r(\Sigma_{i=1}^n u_i) = \hat{r}(\Sigma_{i=1}^n Q_r(u_i))$$

which shows that $Q_{q,r}$ is cone-absolutely summing.

Let r be a continuous seminorm on Y and $T \in L^*(E, Y)$. Then the functional $r \circ T$, defined by

$$(r \circ T)(x) = r(Tx) \quad \text{for all} \quad x \in E$$

is a seminorm on E. The preceding result shows that the quotient map $Q_q : E \to E_q$ is cone-absolutely summing if and only if $\hat{q} \circ Q_q$ is an (PL)-seminorm. This result holds for general case as the following result shows.

(3.2.2) Lemma. <u>Let</u> $(E, C, \mathcal{P})$ <u>be a locally solid space and</u> $T \in L(E, Y)$. <u>Then the following statements are equivalent.</u>

(a) $T \in L^{\ell}(E, Y)$.

(b) $r \circ T$ <u>is an</u> (PL)-<u>seminorm on</u> E <u>for any continuous seminorm</u> r <u>on</u> Y.

(c) <u>For any continuous seminorm</u> r <u>on</u> Y, $r \circ T$ <u>is dominated by some</u> (PL)-<u>seminorm</u> p <u>on</u> E, <u>i.e.</u>,

$$(r \circ T)(x) \leq p(x) \quad \text{for all} \quad x \in E.$$

<u>Proof</u>. The implications (a) ⇒ (b) and (c) ⇒ (a) follow from Proposition (3.1.1), while the implication (b) ⇒ (c) is obvious.

(3.2.3) Lemma. <u>Let</u> $(E, C, \mathcal{P})$ <u>and</u> $(F, K, \mathfrak{L})$ <u>be locally solid spaces and</u> $T \in L^+(E, F)$. <u>If</u> r <u>is an</u> (PL)-<u>seminorm on</u> F, <u>then</u> $r \circ T$ <u>is an</u> (PL)-<u>seminorm on</u> E.

<u>Proof</u>. There exists an $g \in K'$ such that

$$r(y) \leq \sup\{h(y) : h \in [-g, g]\} \quad (y \in F) .$$

As T is positive, $f = T'g \in C'$ and $T'([-g, g]) \subseteq [-f, f]$, hence we have

$$(r \circ T)(x) \leq \sup\{h(Tx) : h \in [-g, g]\} = \sup\{(T'h)(x) : T'h \in [-f, f]\}$$
$$\leq \sup\{\varphi(x) : \varphi \in [-f, f]\} \quad \text{for all} \quad x \in E .$$

This shows that $r \circ T$ is an (PL)-seminorm on E .

(3.2.4) Lemma. Let $(E, C, \mathcal{P})$ be a locally solid space. If p and q are (PL)-seminorms on E , then so do $p + q$ and αp for any $\alpha \geq 0$.

Proof. Clearly αp is an (PL)-seminorm. To prove $p + q$ to be an (PL)-seminorm, let f and g , in C' , be such that

$$p(x) \leq \inf\{f(u) : u \pm x \in C\}$$
$$q(x) \leq \inf\{g(w) : w \pm x \in C\} .$$

Then it is easily seen that

$$(p + q)(x) \leq \inf\{f(v) + g(v) : v \pm x \in C\} \quad (x \in E) ,$$

hence $p + q$ is an (PL)-seminorm.

Recall that a seminorm q on E is additive if

$$q(u + w) = q(u) + q(w) \quad \text{for all} \quad u, w \in C .$$

(3.2.5) Lemma. Let $(E, C, \mathcal{P})$ be a locally solid space, let q be a seminorm on E and let $V = \{x \in E : q(x) \leq 1\}$. Then the following statements are equivalent.

(a) q <u>is</u> $\mathcal{P}$<u>-continuous and additive</u>.

(b) <u>There exists an</u> $f \in C' \cap V^o$ <u>such that</u> $V^o \subset [-f, f]$.

(c) <u>There exists an</u> $f \in C'$ <u>such that</u>

$$q(x) \leq \inf\{f(u) : u \pm x \in C\} \quad \underline{\text{for all}} \quad x \in E$$
$$q(u) = f(u) \quad \underline{\text{for all}} \quad u \in C . \tag{2.2}$$

<u>Proof</u>. In view of Proposition (1.3.15)(b), the seminorm p_{fS} defined by

$$p_{fS}(x) = \inf\{f(u) : u \pm x \in C\}$$

is the gauge of $[-f, f]^o$ and surely $\sigma_S(E, E')$-continuous. If $f \in V^o \cap C'$, then (2.2) holds because $q(u) \leq f(u) \leq q(u)$ for all $u \in C$; if (2.2) holds, then q is additive. Therefore the implications (b) ⇒ (c) ⇒ (a) follow. We complete the proof by showing that (a) implies (b). But this is obvious by Bonsall's theorem (1.1.1) because the additivity of q ensures that q is superlinear on C , hence there exists an $f \in E^*$ such that

$$q(u) \leq f(u) \quad \text{for all} \quad u \in C$$
$$f(x) \leq q(x) \quad \text{for all} \quad x \in E ;$$

it then follows that $f \in C' \cap V^o$ and $V^o \subset [-f, f]$.

As an immediate consequence of Lemmas (3.2.1) and (3.2.5), we obtain:

(3.2.6) Corollary. <u>For a locally solid space</u> $(E, C, \mathcal{P})$, <u>additive continuous seminorms are</u> (PL)-<u>seminorms</u>.

Let X be a locally convex space. Following Schaefer [1], a seminorm q on X is said to be <u>prenuclear</u> if there exists a $\sigma(X', X)$-closed equicontinuous subset B of X' and a positive Radon measure μ on B such that

$$q(x) \leq \int_B |\langle x, x'\rangle| \, d\mu(x') \quad \text{for all} \quad x \in X .$$

Clearly prenuclear seminorms are continuous, and every $\sigma(X, X')$-continuous seminorm on X is prenuclear. Analogues to Lemma (3.2.1) through Lemma (3.2.4) hold for prenuclear seminorms.

(3.2.7) Lemma. Let X be a locally convex space, let q be a seminorm on X and $V = \{x \in X : q(x) \leq 1\}$. Then the following statements are equivalent.

(a) q is a prenuclear seminorm.

(b) The quotient map $Q_q : X \to X_q$ is absolutely summing.

(c) V^o is a prenuclear subset of X' .

(d) There exists a continuous seminorm r on X with $q \leq r$ such that the canonical map $Q_{q,r} : X_r \to X_q$ is absolutely summing.

(e) There exists an absolutely convex o-neighbourhood W in X such that

$$\Sigma_{i=1}^{n} q(x_i) \leq \sup\{\Sigma_{i=1}^{n} |\langle x_i , x'\rangle| : x' \in W^o\}$$

holds for any finite subset $\{x_1, \ldots, x_n\}$ of X .

Proof. As $q = \hat{q} \circ Q_q$, it follows from Proposition (3.1.12) and the definitions that (a), (b) and (e) are equivalent. Since $V^o = Q_q'((Q_q(V))^o)$, it follows from Proposition (3.1.12) that (b) and (c) are equivalent. Clearly the implication (d) $\Rightarrow$ (b) is a consequence of Lemma (3.1.15). Therefore we complete the proof by showing that (e) implies (d) .

Let r be the gauge of W . Then r is a continuous seminorm on X such that $q \leq r$. As $W^o = Q_r'((Q_r(W))^o)$ and $(Q_r(W))^o$ is the polar of the closed unit ball in the normed space X_r , we have, for any finite subset $\{Q_r(x_1), \ldots, Q_r(x_n)\}$ of X_r , that

$$\Sigma_{i=1}^{n} \hat{q}(Q_{q,r}(Q_r(x_i))) = \Sigma_{i=1}^{n} q(x_i) \leq \sup\{\Sigma_{i=1}^{n} |\langle x_i , x'\rangle| : x' \in W^o\}$$

$$= \sup\{\Sigma_{i=1}^{n} |\langle Q_r(x_i), f\rangle| : f \in (Q_r(W))^o\} .$$

Therefore $Q_{q,r}$ is absolutely summing.

(3.2.8) Lemma. _Let_ X _and_ Y _be locally convex spaces and_ $T \in L(X, Y)$. _Then the following statements are equivalent._

(a) $T \in L^s(X, Y)$.

(b) $r \circ T$ _is a prenuclear seminorm on_ X _for any continuous seminorm_ r _on_ Y .

(c) _For any continuous seminorm_ r _on_ Y , $r \circ T$ _is dominated by some prenuclear seminorm_ p _on_ X .

Proof. The implications (a) ⇒ (b) and (c) ⇒ (a) follow from Proposition (3.1.12), while the implication (b) ⇒ (c) is obvious.

(3.2.9) Lemma. _Let_ X _and_ Y _be locally convex spaces and_ $T \in L(X, Y)$. _If_ r _is a prenuclear seminorm on_ Y , _then_ $r \circ T$ _is a prenuclear seminorm on_ X .

Proof. There exists an absolutely convex o-neighbourhood U in Y such that

$$\Sigma_{i=1}^{n} r(y_i) \leq \sup\{\Sigma_{i=1}^{n} |\langle y_i , y'\rangle| : y' \in U^o\}$$

holds for any finite subset $\{y_1, \ldots, y_n\}$ of Y . As T is continuous, there exists an absolutely convex o-neighbourhood W in X such that

$T'(U^o) \subseteq W^o$. For any finite subset $\{x_1, \ldots, x_n\}$ of X , we have that

$$\Sigma_{i=1}^{n}(r \circ T)(x_i) \leq \sup\{\Sigma_{i=1}^{n}|<Tx_i , y'>| : y' \in U^o\}$$
$$\leq \sup\{\Sigma_{i=1}^{n}|<x_i , x'>| : x' \in V^o\} .$$

In view of Lemma (3.2.7), $r \circ T$ is a prenuclear seminorm on X .

(3.2.10) Lemma. <u>Let</u> X <u>be a locally convex space</u>. <u>If</u> p <u>and</u> q <u>are prenuclear seminorms on</u> X , <u>then so do</u> $p + q$ <u>and</u> αp <u>for any</u> $\alpha \geq 0$.

<u>Proof</u>. Clearly αp is prenuclear. To see the prenuclearity of $p + q$, we define $T : X \to X_p \times X_q$ by setting

$$T(x) = (Q_p(x), Q_q(x)) \quad \text{for all} \quad x \in X .$$

Clearly T is linear and $\ker T = \ker(p + q)$, the injection $\hat{T}$ associated with T is an injective linear map from X_{p+q} into $X_p \times X_q$, and

$$T = \hat{T} \circ Q_{p+q} . \tag{2.3}$$

Since $X_p \times X_q$ is a normed space under the norm $\|\cdot\|$ defined by

$$\|(Q_p(x), Q_q(y))\| = p(x) + q(y) \quad \text{for all} \quad x,y \in X$$

and since

$$\|\hat{T}(Q_{p+q}(x))\| = \|Tx\| = p(x) + q(x) = (\widehat{p + q})(Q_{p+q}(x)) ,$$

it follows that $\hat{T}$ is an isometry from X_{p+q} into $X_p \times X_q$. In view of (2.3) and Lemmas (3.1.15) and (3.2.7) , we have only to show that T is absolutely summing.

In fact, by Lemma (3.2.7), there exists absolutely convex o-neighbourhoods V and W such that the inequalities

$$\Sigma_{i=1}^{n} p(x_i) \leq \sup\{\Sigma_{i=1}^{n} |\langle x_i , x'\rangle| : x' \in V^o\}$$

$$\Sigma_{j=1}^{m} q(z_j) \leq \sup\{\Sigma_{j=1}^{m} |\langle z_j , z'\rangle| : z' \in W^o\}$$

hold for finite subsets $\{x_1, \ldots, x_n\}$ and $\{z_1, \ldots, z_m\}$ of X respectively. Setting

$$U = 2^{-1}(V \cap W) .$$

Then U is an absolutely convex o-neighbourhood in X, and V^o, W^o are subsets of $(V \cap W)^o$. For any subset $\{x_1, \ldots, x_n\}$ of X, we have that

$$\begin{aligned}
\Sigma_{i=1}^{n} \|Tx_i\| &= \Sigma_{i=1}^{n} (p(x_i) + q(x_i)) \\
&\leq \sup\{\Sigma_{i=1}^{n} |\langle x_i , x'\rangle| : x' \in V^o\} + \sup\{\Sigma_{i=1}^{n} |\langle x_i , z'\rangle| : z' \in W^o\} \\
&\leq \sup\{\Sigma_{i=1}^{n} |\langle x_i , 2f\rangle| : f \in (V \cap W)^o\} \\
&= \sup\{\Sigma_{i=1}^{n} |\langle x_i , g\rangle| : g \in 2(V \cap W)^o\} \\
&= \sup\{\Sigma_{i=1}^{n} |\langle x_i , g\rangle| : g \in U^o\} .
\end{aligned}$$

Therefore T is absolutely summing.

Let Δ denote the Fréchet space of rapidly decreasing sequences, namely the vector space of all numbers sequences $\lambda = (\lambda_n)$ such that for any integer k

$$q_k(\lambda) = \Sigma_n n^k |\lambda_n| < \infty ,$$

which is equipped with the topology determined by the family $\{q_k\}$ of seminorms. A seminorm q on X is said to be <u>strongly nuclear</u> (resp. <u>quasi-nuclear</u>) if there exists $\lambda = (\lambda_n)$ in Δ (resp. in ℓ^1) and an equicontinuous sequence (f_n) in X' such that

$$q(x) \leq \Sigma_n |\lambda_n \langle x, f_n\rangle| \quad \text{for all } x \in X .$$

Clearly each $\sigma(X, X')$-continuous seminorm on X is quasi-nuclear, and strongly nuclear seminorms are quasi-nuclear. Furthermore, it is easily verified that quasi-nuclear seminorms on X are prenuclear.

In view of Lemmas (3.2.1), (3.2.7) and (3.1.16), we obtain:

(3.2.11) Lemma. *Let* $(E, C, \mathscr{P})$ *be a locally solid space. Then prenuclear seminorms on* E *are* (PL)-*seminorms*.

We shall give an example (see Examples (3.2.16)(b)) to show that the converse of the preceding result need not be valid.

In terms of cone-absolutely summing maps we present below some characterizations of a locally solid topology to be the topology of uniform convergence on all order-intervals.

(3.2.12) Theorem. *For a locally solid space* $(E, C, \mathscr{P})$, *the following statements are equivalent*.

(a) $\mathscr{P} = \sigma_S(E, E')$ (or $o(E, E')$) .

(b) *Every* $\mathscr{P}$-*continuous seminorm on* E *is an* (PL)-*seminorm*.

(c) *Every* $\mathscr{P}$-*equicontinuous subset of* E' *is order-bounded*.

(d) $Q_q \in L^{\ell}(E, E_q)$ *for any continuous seminorm* q *on* E .

(e) *For any continuous seminorm* q *on* E *there exists a continuous seminorm* r *on* E *with* $q \leq r$ *such that* $Q_{q,r} \in L^{\ell}(E_r, E_q)$.

(f) *The identity map* I *belongs to* $L^{\ell}(E, E)$.

(g) *The canonical embedding map of* $(\ell^1_a\langle E\rangle, C_\pi(E), \mathscr{P}_{\pi D})$ *into* $(\ell^1\langle E\rangle, C_\varepsilon(E), \mathscr{P}_{\varepsilon D})$ *is a topological isomorphism from the first space onto the second*.

(h) $L^{\ell}(E, Y) = L(E, Y)$ *for any locally convex space* Y .

(i) $L^{\ell}(E, Y) = L(E, Y)$ *for any normed space* Y .

Moreover, if in addition, $(E, C, \mathcal{P})$ is metrizable (resp. Fréchet) space, then $\mathcal{P} = \sigma_S(E, E')$ if and only if the canonical embedding map of $(\ell^1[E], C_\pi(E), \mathcal{P}_\pi)$ into $(\ell^1\langle E\rangle, C_\varepsilon(E), \mathcal{P}_{\varepsilon D})$ is a topological isomorphism from the first space onto the second (resp. each positive summable sequence in E is absolutely summable).

Proof. In view of Lemma (3.2.1) and Corollary (3.1.4), the statements (a) through (g) are equivalent, and the last part also holds. The implication (h) $\Rightarrow$ (f) and (h) $\Rightarrow$ (i) $\Rightarrow$ (d) are obvious, while the implication (f) $\Rightarrow$ (i) is a consequence of Lemma (3.1.5) because I is positive.

Let $(E, C, \mathcal{P})$ be a locally solid space. If $\mathcal{P}$ is determined by a family of additive seminorms, then $\mathcal{P} = \sigma_S(E, E')$ in view of Theorem (3.2.12) and Corollary (3.2.6). In particular, the topology on a base normed space $(E, C, \|\cdot\|)$ is $\sigma_S(E, E')$.

A locally convex space $(X, \mathcal{P})$ is called a nuclear space if every $\mathcal{P}$-continuous seminorm on X is prenuclear.

In terms of absolutely summing maps we present some characterizations of locally convex spaces to be nuclear, which is analog to Theorem (3.2.10).

(3.2.13) Theorem. For a locally convex space $(X, \mathcal{P})$, the following statements are equivalent.

(a) $(X, \mathcal{P})$ is a nuclear space.

(b) Every $\mathcal{P}$-equicontinuous subset of X' is prenuclear.

(c) $Q_q \in L^s(X, X_q)$ for any continuous seminorm q on X .

(d) For any continuous seminorm q on X there exists a continuous seminorm r on X with $q \leq r$ such that $Q_{q,r} \in L^s(X_r, X_q)$.

(e) <u>The identity map</u> I <u>belongs to</u> $L^s(X, X)$.

(f) <u>The canonical embedding map of</u> $(\ell^1[X], \mathcal{P}_\pi)$ <u>into</u> $(\ell^1(X), \mathcal{P}_\varepsilon)$ <u>is a topological isomorphism from the first space onto the second</u>.

(g) $L^s(X, Y) = L(X, Y)$ <u>for any locally convex space</u> Y .

(h) $L^s(X, Y) = L(X, Y)$ <u>for any normed space</u> Y .

<u>Moreover, if in addition,</u> $(X, \mathcal{P})$ <u>is metrizable, then</u> X <u>is nuclear if and only if every summable sequence in</u> X <u>is absolutely summable</u>.

<u>Proof</u>. In view of Lemma (3.2.7) and Corollary (3.1.14), the statements (a) through (f) are equivalent, and the last part also holds. The implications (g) ⇒ (e) and (g) ⇒ (h) ⇒ (c) are obvious, while the implication (e) ⇒ (g) is a consequence of Lemma (3.1.15).

The equivalence of (a) and (f), due to Pietsch, is an important criterion of nuclearity.

The preceding two theorems have many important applications, we mention a few below.

(3.2.14) Corollary. <u>A locally solid space</u> $(E, C, \mathcal{P})$ <u>is nuclear if and only if it satisfies the following two conditions</u>:

(i) $\mathcal{P} = \sigma_S(E, E')$;

(ii) <u>every order-bounded subset of</u> E' <u>is prenuclear</u>.

<u>Proof</u>. The necessity follows from Theorems (3.2.12) and (3.2.13) and Lemma (3.1.16). Conversely, if E satisfies conditions (i) and (ii), then $\{[-f, f] : f \in C'\}$ is a fundamental system of equicontinuous sets in E' , and hence the condition (ii) implies that each $\mathcal{P}$-equicontinuous subset

of E' is prenuclear; consequently E is a nuclear space in view of Theorem (3.2.13).

The necessity of the preceding result is a generalization of Kōmura and Koshi's result [1] .

(3.2.15) Corollary. A Fréchet locally solid space $(E, C, \mathscr{P})$ is nuclear if and only if it satisfies the following two conditions:

(i) $\mathscr{P} = \sigma_S(E, E')$;

(ii) every summable sequence in E is the difference of two positive summable sequences in E .

Proof. Follows from Theorems (3.2.12) and (3.2.13).

The norm topology on ℓ^1 coincides with $\sigma_S(\ell^1, \ell^\infty)$, but ℓ^1 is not nuclear, therefore condition (ii) in Corollary (3.2.15) is essential.

(3.2.16) Examples.

(a) Cone-absolutely summing maps need not be absolutely summing.

The norm topology on ℓ^1 coincides with $\sigma_S(\ell^1, \ell^\infty)$, but ℓ^1 is not nuclear, hence in view of Theorems (3.2.12) and (3.2.13), $I \in L^\ell(\ell^1, \ell^1)$ but $I \notin L^s(\ell^1, \ell^1)$.

(b) Continuous linear maps need not be cone-absolutely summing.

It is well-known that the space c_o consisting of all null-sequences of real numbers is a Banach lattice equipped with the usual norm and usual ordering, and that the norm topology is strictly finer than $\sigma_S(c_o, \ell^1)$. It follows from Theorem (3.2.12) that $I \notin L^\ell(c_o, c_o)$, but $I \in L(c_o, c_o)$; also the usual norm on c_o is not an (PL)-norm.

We can deal with Schwartz spaces in the same manner. We conclude this section with a criterion of locally convex spaces to be Schwartz spaces.

Recall that an $T \in L^*(X, Y)$ is called a <u>precompact</u> map if it sends some o-neighbourhood in X into a precompact subset of Y. Clearly every precompact map must be continuous. The set consisting of all precompact maps from X into Y, denoted by $L^p(X, Y)$, is a vector subspace of $L(X, Y)$. It is easily seen that if $T \in L(X, Y)$, if $S \in L(Y, Z)$ and if one of T and S is precompact, then $S \circ T \in L^p(X, Z)$.

A seminorm q on X is said to be <u>precompact</u> if there exists $(\lambda_n) \in c_o$ and an equicontinuous sequence (f_n) in X' such that

$$q(x) \leq \sup_n |\lambda_n f_n(x)| \quad \text{for all} \quad x \in X .$$

Clearly precompact seminorms on X are continuous, the sum of two precompact seminorms and the positive scalar product of a precompact seminorm are precompact. Terzioǧlu [1] and Randtke [1] have shown that a seminorm q on X is precompact if and only if $Q_q \in L^p(X, Y_q)$.

A locally convex space X is called a <u>Schwartz space</u> if every continuous seminorm on X is precompact.

Nuclear spaces are Schwartz spaces, but the converse need not be valid. Analog to Theorem (3.2.13) holds for Schwartz spaces, we mention here, but the proof is omitted.

For a locally convex space X, the following statement are equivalent:

(a) X <u>is a Schwartz space</u>;

(b) $Q_q \in L^p(X, X_q)$ <u>for any continuous seminorm</u> q <u>on</u> X ;

(c) <u>for any continuous seminorm</u> q <u>on</u> X <u>there exists a continuous seminorm</u> r <u>on</u> X <u>with</u> $q \leq r$ <u>such that</u> $Q_{q,r} \in L^p(X_r, X_q)$;

(d) $I \in L^p(X, X)$;

(e) $L^p(X, Y) = L(X, Y)$ <u>for any locally convex space</u> Y ;

(f) $L^p(X, Y) = L(X, Y)$ <u>for any normed space</u> Y .

For further information on Schwartz spaces, we refer the reader to Grothendieck [1] and Horváth [1] .

3.3 Cone-prenuclear mappings

Recall that a linear map T from a locally convex space X into another locally convex space Y is <u>bounded</u> if it sends some o-neighbourhood in X into a bounded subset of Y , and if and only if there exists a continuous seminorm p on X with the following property: for any continuous seminorm q on Y there is $\alpha_q \geq 0$ for which the inequality

$$q(Tx) \leq \alpha_q p(x) \quad \text{for all} \quad x \in X \tag{3.1}$$

holds; it then follows that every bounded linear map must be continuous. If, in addition, Y is a normed space, then every continuous linear map from X into Y is bounded, we shall see from Examples (3.3.14)(2) that the assumption of normability of Y is essential. The set consisting of all bounded linear maps from X into Y , denoted by $L^{\ell b}(X, Y)$, is a vector subspace of $L(X, Y)$, and

$$L^p(X, Y) \subset L^{\ell b}(X, Y) ,$$

where $L^p(X, Y)$ is the vector space consisting of all precompact maps from

X into Y. As bounded subsets of Y are $\sigma(Y, Y')$-precompact, it follows that

$$L^{\ell b}(X, Y) = L^{p}(X, Y(\sigma)) .$$

Let $(E, C, \mathcal{P})$ be an ordered convex space. A linear map T from E into Y is called a <u>cone-prenuclear map</u> if there exists an (PL)-seminorm p on E such that the set $\{Tx : p(x) \leq 1\}$ is bounded in Y. Clearly, T is cone-prenuclear if and only if there exists an (PL)-seminorm p on E with the following property: for any continuous seminorm q on Y, there is $\alpha_q \geq 0$ for which the inequality

$$q(Tx) \leq \alpha_q p(x) \tag{3.2}$$

holds for all $x \in E$.

If $(E, C, \mathcal{P})$ is a locally solid space, then every (PL)-seminorm on E is continuous, hence every cone-prenuclear map must be bounded and <u>a fortiori</u> continuous in view of Formulae (3.1) and (3.2). The set consisting of all cone-prenuclear maps from E into Y, denoted by $L^{\ell n}(E, Y)$, is a vector subspace of $L^{\ell b}(E, Y)$ by Lemma (3.2.4). As p is an (PL)-seminorm, it follows from Lemma (3.2.2) and Formula (3.2) that

$$L^{\ell n}(E, Y) \subseteq L^{\ell}(E, Y) \cap L^{\ell b}(E, Y) .$$

If, in addition, Y is a normed space, then

$$L^{\ell n}(E, Y) = L^{\ell}(E, Y) .$$

Furthermore, we have the following criteria for the topology $\sigma_S(E, E')$.

(3.3.1) Theorem. <u>Let</u> $(E, C, \mathcal{P})$ <u>be a locally solid space</u>. <u>For any locally convex space</u> Y, <u>the following statements are equivalent</u>.

(1) $\mathscr{P} = \sigma_S(E, E')$.

(2) $L^{\ell n}(E, Y) = L^{\ell b}(E, Y)$.

(3) $L^{\ell b}(E, Y) \subseteq L^{\ell}(E, Y)$.

(4) $L^{p}(E, Y) \subseteq L^{\ell n}(E, Y)$.

<u>Proof</u>. As $L^{\ell n}(E, Y) \subseteq L^{\ell}(E, Y)$, the implication (2) ⇒ (3) follows, while the implication (2) ⇒ (4) is an immediate consequence of the fact that $L^{p}(E, Y) \subseteq L^{\ell b}(E, Y)$. In view of the definitions of bounded linear maps and cone-prenuclear maps, (1) implies (2). Therefore we complete the proof by showing that (3) ⇒ (1) and (4) ⇒ (1) .

(3) ⇒ (1) : If the statement (3) holds for any locally convex space Y , then (3) holds for any normed space Y . For any normed space Y , we have that $L^{\ell b}(E, Y) = L(E, Y)$, and hence $L(E, Y) \subseteq L^{\ell}(E, Y)$ for any normed space Y . In view of Theorem (3.2.12), (3) implies (1) .

(4) ⇒ (1) : It is sufficient to show, in view of Theorem (3.2.12), that $L(E, Y) \subseteq L^{\ell n}(E, Y)$ holds for any normed space Y . In fact, for a normed space Y , we have that

$$L(E, Y) = L^{\ell b}(E, Y) . \tag{3.3}$$

Note that the following hold for any locally convex space Y

$$L^{\ell b}(E, Y) = L^{p}(E, Y(\sigma)) \quad \text{and} \quad L^{\ell n}(E, Y) = L^{\ell n}(E, Y(\sigma)) . \tag{3.4}$$

We conclude from Formulae (3.3), (3.4) and the assumption that

$$L(E, Y) = L^{\ell b}(E, Y) = L^{p}(E, Y(\sigma)) \subseteq L^{\ell n}(E, Y(\sigma)) = L^{\ell n}(E, Y) ,$$

which obtains our required assertion.

As a consequence of Lemmas (3.2.2) and (3.2.3), we obtains:

(3.3.2) Lemma. Let E, F be locally solid spaces, and let X , Y be locally convex spaces. Then the following statements hold:

(1) If $T \in L^{\ell n}(E, X)$ and $S \in L(X, Y)$, then $S \circ T \in L^{\ell n}(E, Y)$.

(2) If $S \in L(E, F)$ is positive and if $T \in L^{\ell n}(F, Y)$, then $T \circ S \in L^{\ell n}(E, Y)$.

Let $T \in L(E, Y)$, let V be a subset of E and let M be a subset of Y' . Then it is easily shown that

$$(T(V))^{\circ} = (T')^{-1}(V^{\circ}) \quad \text{and} \quad (T'(M))^{\circ} = T^{-1}(M^{\circ}) . \tag{3.5}$$

(3.3.3) Proposition. Let $(E, C, \mathcal{P})$ be a locally solid space, let Y be a locally convex space and let $T \in L(E, Y)$. Then the following statements are equivalent.

(a) $T \in L^{\ell n}(E, Y)$.

(b) There exists an $f \in C'$ such that for any o-neighbourhood U in Y , there is an $\alpha_U \geq 0$ for which $T'(U^{\circ}) \subseteq \alpha_U[-f, f]$.

(c) T' sends some $\beta(Y', Y)$-neighbourhood of 0 in Y' into an order-bounded subset of E' , i.e., $T' \in L^{o}(Y'(\beta), E')$.

(d) T is the compose of the following three continuous linear maps

$$E \xrightarrow{Q} G \xrightarrow{\hat{T}} H \xrightarrow{J} Y$$

where G is a base normed space, H is a normed space, $Q \in L(E, G)$ is positive, $\hat{T} \in L^{\ell}(G, H)$, and J is a continuous linear map.

Proof. The equivalence of (a) and (b) is a consequence of Lemma

(3.2.1), the equivalence of (a) and (c) follows from Formula (3.5), and the implication (d) $\Rightarrow$ (a) is a consequence of Lemma (3.3.2) on account of $L^{\ell}(G, H) = L^{\ell n}(G, H)$. It remains to verify that (a) implies (d) .

Let p be an (PL)-seminorm on E such that $B = \{Tx : p(x) \leq 1\}$ is a bounded subset of Y . Then the boundedness of B insures that $\mathrm{Ker}\, p \subseteq \mathrm{Ker}\, T$, hence there exists a continuous linear map S from E_p onto $Y(B)$ (since $\|S\| \leq 1$) such that $T = J_B \circ S \circ Q_p$. On the other hand, as p is an (PL)-seminorm, there exists an $h \in C'$ such that

$$p(x) \leq \inf\{h(u) : u \pm x \in C\} \quad \text{for any } x \in E .$$

The seminorm r defined by

$$r(x) = \inf\{h(u) : u \pm x \in C\} \quad \text{for any } x \in E ,$$

is additive on C for which $Q_{p,r} : E_r \to E_p$ is cone-absolutely summing, thus the map $\hat{T}$ defined by

$$\hat{T} = S \circ Q_{p,r}$$

is a cone-absolutely summing map from the normed space E_r onto the normed space $Y(B)$ and satisfies

$$T = J_B \circ \hat{T} \circ Q_r .$$

Clearly Q_r is positive. We complete the proof by showing that E_r is a base normed space.

In fact, for any $Q_r(u) \in Q_r(C)$, we can assume without loss of generality that $u \in C$. If $Q_r(w)$ and $Q_r(u)$ belong to $Q_r(C)$, then we have

$$\hat{r}(Q_r(u) + Q_r(w)) = r(u + w) = h(u + w) = \hat{r}(Q_r(u)) + \hat{r}(Q_r(w))$$

which shows that $\hat{r}$ is additive on $Q_r(C)$. On the other hand, if $\hat{r}(Q_r(x)) < 1$, then by the definition of r, there exists $u \in C$ such that

$$u \pm x \in C \quad \text{and} \quad h(u) < 1 .$$

As $Q_r(u) \pm Q_r(x) \in Q_r(C)$ and $\hat{r}(Q_r(u)) = h(u) < 1$, it follows that the open unit ball Σ in E_r is absolutely dominated, and hence that Σ is solid. Thus E_r is a base normed space (see Wong and Ng [1, (9.5)]).

(3.3.4) Corollary. Let $(E, C, \mathcal{P})$ be a locally solid space. Then the identity map on E is cone-prenuclear if and only if E' has an order-unit and $\mathcal{P} = \sigma_S(E, E')$.

Proof. In view of Theorem (3.2.12), $\mathcal{P} = \sigma_S(E, E')$ if and only if each $\mathcal{P}$-equicontinuous subset of E' is order-bounded. The result then follows immediately from Proposition (3.3.3).

(3.3.5) Corollary. Let $(E, C, \|\cdot\|)$ be a Banach space which is locally solid. If E' does not have any order-unit, then there exists a positive summable sequence in E which is not absolutely summable.

Proof. Suppose that every positive summable sequence in E is absolutely summable. Then by Theorem (3.2.12), the identity map I on E is cone-absolutely summing, and hence I is cone-prenuclear by the normability of E. Consequently E' has an order-unit by the preceding corollary.

The following result is concerning the lattice properties of cone-prenuclear linear maps.

(3.3.6) Proposition. Let $(G, C, \mathcal{P})$ and $(H, K, \mathcal{I})$ be locally convex Riesz spaces. If $(H, K, \mathcal{I})$ is both locally and boundedly order complete, then $L^{\ell n}(G, H)$ is an ℓ-ideal in $L^b(G, H)$.

Proof. As $L^{\ell n}(G, H) \subset L^{\ell}(G, H)$, it follows from Theorem (3.1.6) that $|T| \in L^{\ell}(G, H)$ for all $T \in L^{\ell n}(G, H)$. We now show that $|T| \in L^{\ell n}(G, H)$.

In fact, there exists $f \in C'$ such that for any continuous Riesz seminorm q on H there is $\alpha_q \geq 0$ for which the inequality

$$q(Tx) \leq \alpha_q \langle |x|, f\rangle \tag{3.6}$$

holds for all $x \in G$. Since $(H, K, \mathcal{I})$ is locally order complete, it follows from the proof of Theorem (3.1.6) that there exists a continuous Riesz seminorm q on H such that the inequality

$$r(\sup B) \leq \sup\{q(b) : b \in B\}$$

holds for any bounded subset B of H which is directed upwards. For any $u \in C$, the set $\{\Sigma_{i=1}^n |Tu_i| : u_i \in C , \Sigma_{i=1}^n u_i = u\}$ is bounded in H and directed upwards, also

$$|T|(u) = \sup\{\Sigma_{i=1}^n |Tu_i| : u_i \in C , \Sigma_{i=1}^n u_i = u\} ,$$

then we have that

$$r(|T|u) \leq \sup\{q(\Sigma_{i=1}^n |Tu_i|) : u_i \in C , \Sigma_{i=1}^n u_i = u\} \quad (u \in C) .$$

For this q , there exists $\alpha_q \geq 0$ such that (3.6) holds for all $x \in E$, hence for any finite subset $\{u_1, \ldots, u_n\}$ of C with $\Sigma_{i=1}^n u_i = u$, we obtain

$$q(\Sigma_{i=1}^n |Tu_i|) \leq \Sigma_{i=1}^n q(Tu_i) \leq \alpha_q \langle\Sigma_{i=1}^n u_i, f\rangle = \alpha_q\langle u, f\rangle ,$$

consequently,

$$r(|T|u) \leq \alpha_q \langle u, f\rangle \quad \text{for all} \quad u \in C .$$

Now for any $x \in G$, we have

$$r(|T|x) \leq r(|T|(|x|)) \leq \alpha_q \langle |x|, f\rangle$$

which implies that $|T| \in L^{\ell n}(G, H)$.

Finally, it is not hard to see that $L^{\ell n}(G, H)$ is a solid subspace of $L^b(G, H)$. Therefore the proof is complete.

(3.3.7) Corollary. Let $(G, C, \mathcal{P})$ and $(H, K, \mathcal{L})$ be locally convex Riesz spaces. If $\mathcal{P} = \sigma_S(G, G')$ and if $(H, K, \mathcal{L})$ is both locally and boundedly order complete, then $L^{\ell b}(G, H)$ is an ℓ-ideal in $L^b(G, H)$.

Proof. This follows from Theorem (3.3.1) and Proposition (3.3.6).

Let X and Y be locally convex spaces. A linear map $T : X \to Y$ is said to be prenuclear (resp. quasi-nuclear) if there exists a prenuclear (resp. quasi-nuclear) seminorm p on X such that $\{Tx : p(x) \leq 1\}$ is a bounded subset of Y . Clearly, T is prenuclear (resp. quasi-nuclear) if and only if there exists a prenuclear (resp. quasi-nuclear) seminorm p on X with the following property: for any continuous seminorm q on Y , there is $\alpha_q \geq 0$ for which the inequality

$$q(Tx) \leq \alpha_q \, p(x) \tag{3.7}$$

holds for all $x \in X$. Consequently prenuclear maps and quasi-nuclear maps are continuous. In view of Lemma (3.2.10) and Formulae (3.7) and (3.1), the set consisting of all prenuclear (resp. quasi-nuclear) maps from X into Y , denoted by $L^{pn}(X, Y)$ (resp. $L^{qn}(X, Y)$), is a vector subspace of $L^{\ell b}(X, Y)$. As each quasi-nuclear seminorm is prenuclear, it follows from Lemma (3.2.8) that

$$L^{qn}(X, Y) \subseteq L^{pn}(X, Y) \subseteq L^{s}(X, Y) \cap L^{\ell b}(X, Y) .$$

If, in addition, Y is a normed space, then

$$L^{pn}(X, Y) = L^{s}(X, Y) .$$

On the other hand, if X and Y are normed spaces, then our definition of quasi-nuclear maps coincides with the usual one defined by Pietsch [1], hence the example constructed by Pietsch [1, p.44 and 60] shows that $L^{qn}(X, Y)$ is, in general, a <u>proper</u> vector subspace of $L^{pn}(X, Y)$.

If E is a locally solid space, then Lemma (3.2.11) ensures that

$$L^{pn}(E, Y) \subseteq L^{\ell n}(E, Y) .$$

It is amusement to compare the following results with Corollaries (3.3.4) and (3.3.5)

(3.3.8) Proposition. <u>Let</u> X <u>be a locally convex space</u>. <u>Then the identity map on</u> X <u>is prenuclear if and only if</u> X <u>is normable and finite dimensional</u>.

<u>Proof</u>. Suppose that the identity map on X is prenuclear. Then there exists a prenuclear seminorm p on X with the following property: for any continuous seminorm q on X there is $\alpha_q \geq 0$ for which

$$q(x) \leq \alpha_q\, p(x) \quad \text{for all} \quad x \in X .$$

Hence the topology on X is determined by the single prenuclear seminorm p .

It then follows from Theorem (3.2.13) that X is a nuclear normed space, and hence that X is a finite dimensional normed space.

Conversely, if X is normable and finite dimensional, then X is nuclear, and hence the identity map I on X is absolutely summing by Theorem (3.2.13) , consequently I is prenuclear because X is a normed space.

As an application of the Proposition (3.3.8) and Theorem (3.2.13), we obtain the following famous theorem of Dvoretzky and Rogers.

(3.3.9) Corollary (Droretzky and Rogers). _A normed space_ X _is finite dimensional if and only if every summable sequence in_ X _is absolutely summable_.

We now present other characterizations for nuclear spaces, which is anologues to Theorem (3.3.1), by means of prenuclear maps as follows.

(3.3.10) Theorem. _Let_ X _be a locally convex space_. _The following statements are equivalent_.

(1) X _is a nuclear space_.

(2) $L^{pn}(X, Y) = L^{\ell b}(X, Y)$ _for any locally convex space_ Y .

(3) $L^{\ell b}(X, Y) \subseteq L^{s}(X, Y)$ _for any locally convex space_ Y .

(4) $L^{p}(X, Y) \subseteq L^{pn}(X, Y)$ _for any locally convex space_ Y .

Proof. As $L^{pn}(X, Y) \subseteq L^{s}(X, Y)$, the implication (2) ⇒ (3) follows, while the implication (2) ⇒ (4) is obvious. In view of Theorem (3.2.13) and the definition of prenuclear maps, (1) implies (2). The proofs of (3) ⇒ (1) and (4) ⇒ (1) are similar to that given in Theorem (3.3.1), and hence will be omitted.

An absolutely convex bounded subset B of Y is said to be <u>infracomplete</u> if the normed space Y(B) is complete. It is easily seen that every absolutely convex bounded which is sequentially complete in itself is infracomplete.

Recall that a continuous linear map $T : X \to Y$ is <u>nuclear</u> if it is of the form

$$T(x) = \Sigma_n \lambda_n f_n(x) y_n \quad \text{for all} \quad x \in X , \tag{3.8}$$

where $(\lambda_n) \in \ell^1$, (f_n) is an equicontinuous sequence in X' , and (y_n) is a sequence contained in some infracomplete subset B of Y . Formula (3.8) is referred to as a <u>nuclear representation of</u> T . The set consisting of all nuclear maps from X into Y , denoted by $L^n(X, Y)$, is a vector subspace of L(X, Y) . A locally convex space X is nuclear if and only if $L^n(X, Y) = L(X, Y)$ for any Banach space Y (see Schaefer [1, p.60]).

Suppose now that $T \in L^n(X, Y)$ and that T has a nuclear representation (3.8). Then the functional p , defined by

$$p(x) = \Sigma_n |\lambda_n| \, |f_n(x)| \quad \text{for any} \quad x \in X \tag{3.9}$$

is a quasi-nuclear seminorm on X since the series $\Sigma_n |\lambda_n| \, |f_n(x)|$ converges for any $x \in X$. For any continuous seminorm q on Y , the boundedness of (y_n) ensures that there exists an $\alpha_q \geq 0$ such that

$$q(y_n) \leq \alpha_q \quad \text{for all} \quad n \geq 0 ,$$

it then follows from (3.8) and (3.9) that

$$q(Tx) \leq \alpha_q \, p(x) \quad \text{for all} \quad x \in X .$$

Therefore T is a quasi-nuclear map, thus

$$L^n(X, Y) \subseteq L^{qn}(X, Y) \subseteq L^{pn}(X, Y) \subseteq L^s(X, Y) \cap L^{\ell b}(X, Y) .$$

We shall see from Corollary (3.3.13) that the composite of two prenuclear maps, <u>and a fortiori</u> of quasi-nuclear maps, is nuclear.

Analogues to (3.3.2) and (3.3.3) holds for prenuclear maps as the following two results show.

(3.3.11) Lemma. <u>Let</u> $T \in L(X, Y)$ <u>and</u> $S \in L(Y, Z)$, <u>where</u> X, Y <u>and</u> Z <u>are locally convex spaces</u>. <u>If one of</u> T <u>and</u> S <u>is prenuclear</u> (<u>resp</u>. <u>quasi-nuclear</u>, <u>nuclear</u>), <u>then</u> $S \circ T$ <u>is a prenuclear</u> (<u>resp</u>. <u>quasi-nuclear</u>, <u>nuclear</u>) <u>map</u>.

<u>Proof</u>. This is an immediate consequence of Lemmas(3.2.8) and (3.2.9).

(3.3.12) Proposition. <u>Let</u> X <u>and</u> Y <u>be locally convex spaces and suppose that</u> $T \in L(X, Y)$. <u>Then the following statements are equivalent</u>.

(a) $T \in L^{pn}(X, Y)$.

(b) <u>There exists an</u> $\sigma(X', X)$-<u>closed equicontinuous subset</u> B <u>of</u> X' <u>and a positive Radon measure</u> μ <u>on</u> B <u>such that for any continuous seminorm</u> q <u>on</u> Y, <u>there is</u> $\alpha_q \geq 0$ <u>for which the inequality</u>

$$q(Tx) \leq \int_B |\langle x, x'\rangle| d\mu(x')$$

holds for all $x \in X$.

(c) <u>There exists an absolutely convex</u> o-<u>neighbourhood</u> W <u>in</u> X <u>such that for any continuous seminorm</u> q <u>on</u> Y <u>there is</u> $\beta_q \geq 0$ <u>for which the inequality</u>

$$\Sigma_{i=1}^{n} q(Tx_i) \leq \beta_q \sup\{\Sigma_{i=1}^{n} |\langle x_i, x'\rangle| : x' \in W^o\}$$

holds for any finite subset $\{x_1, \dots, x_n\}$ of X .

(d) T is the compose of the following three continuous linear maps

$$X \xrightarrow{Q} Z \xrightarrow{\hat{T}} M \xrightarrow{J} Y$$

where Z and M are normed spaces and $\hat{T} \in L^s(Z, M)$.

Proof. In view of (3.7), the equivalence of (a) through (c) follows from Lemma (3.2.7), while the proof of the equivalence of (a) and (b) is similar to that given in Proposition (3.3.3), and hence will be omitted.

(3.3.13) Corollary. The composite of two prenuclear maps, and surely of quasi-nuclear maps, is nuclear.

Proof. Let $T \in L^{pn}(X, Y)$ and $S \in L^{pn}(Y, Z)$, where X, Y and Z are locally convex spaces. By Proposition (3.3.12), T and S can be decomposed in the following way:

$$X \xrightarrow{Q_1} X_1 \xrightarrow{\hat{T}} X_2 \xrightarrow{J_1} Y$$

$$Y \xrightarrow{Q_2} Y_1 \xrightarrow{\hat{S}} Y_2 \xrightarrow{J_2} Z$$

where X_i and Y_i $(i = 1, 2)$ are normed spaces, $\hat{T} \in L^s(X_1, X_2)$ and $\hat{S} \in L^s(Y_1, Y_2)$. Define

$$R = Q_2 \circ J_1 \circ \hat{T} .$$

Then $R \in L^s(X_1, Y_1)$ by Lemma (3.1.15), hence $\hat{S} \circ R$ is a nuclear map from X_1 into Y_2 by Pietsch [1, (3.3.5)] because X_1, Y_1 and Y_2 are normed spaces. As

$$S \circ T = J_2 \circ \hat{S} \circ R \circ Q_1 ,$$

we conclude from Lemma (3.3.12) that $S \circ T$ is nuclear.

Finally, if S and T are quasi-nuclear, then they must be prenuclear, and hence $S \circ T$ is nuclear by the above conclusion.

(3.3.14) Examples

(1) Cone-absolutely summing maps are neither cone-prenuclear nor precompact.

It is known that ℓ^1 is the Banach dual of the Banach lattice c_o, and that ℓ^1 does not have any order-unit. It follows from Theorem (3.2.12) and Corollary (3.3.4) that

$$I \in L^{\ell}(c_o(\sigma_S)) \quad \text{and} \quad I \notin L^{\ell n}(c_o(\sigma_S)) ,$$

and hence from Theorem (3.3.1) that

$$I \notin L^{p}(c_o(\sigma_S)) .$$

(2) Cone-absolutely summing maps (in particular, continuous linear maps) need not be bounded, and bounded linear maps need not be cone-prenuclear.

As $L^{\ell n}(E, Y) \subset L^{\ell b}(E, Y)$ hold always, it follows from the preceding example that

$$I \in L^{\ell}(c_o(\sigma_S)) \quad \text{and} \quad I \notin L^{\ell b}(c_o(\sigma_S)) .$$

On the other hand, the norm-topology $\|.\|$ on c_o is strictly finer than $\sigma_S(c_o, \ell^1)$, it follows from Theorem (3.2.12) that

$$I \in L^{\ell b}(c_o) \quad \text{and} \quad I \notin L^{\ell n}(c_o)$$

because of $L^{\ell b}(c_o) = L(c_o)$ and $L^{\ell n}(c_o) = L^{\ell}(c_o)$.

(3) Absolutely summing maps are neither prenuclear nor precompact nor nuclear.

For any infinite dimensional locally convex space X, it is trivial that $X(\sigma) = (X, \sigma(X, X'))$ is a nuclear space. We have, by Theorem (3.2.13) and Proposition (3.3.8), that

$$I \in L^{s}(X(\sigma)) \quad \text{and} \quad I \notin L^{pn}(X(\sigma)) ,$$

and hence from Theorem (3.3.10) and the fact that $L^{n}(X, Y) \subset L^{pn}(X, Y)$,

$$I \notin L^{p}(X(\sigma)) \quad \text{and} \quad I \notin L^{n}(X(\sigma)) .$$

(4) Cone-absolutely summing maps need not be absolutely summing.

It is known that the norm-topology on ℓ^1 coincides with $\sigma_S(\ell^1, \ell^\infty)$, and that ℓ^1 is infinite dimensional. By Theorem (3.2.12) and (3.2.13),

$$I \in L^{\ell}(\ell^1) \quad \text{and} \quad I \notin L^{s}(\ell^1) .$$

(5) Cone-prenuclear maps need not be prenuclear.

Since ℓ^1 is a Banach space, it follows that

$$L^{\ell}(\ell^1) = L^{\ell n}(\ell^1) \quad \text{and} \quad L^{s}(\ell^1) = L^{pn}(\ell^1) .$$

Therefore $I \in L^{\ell n}(\ell^1)$ but $I \notin L^{pn}(\ell^1)$.

BIBLIOGRAPHY

This bibliography covers the works directly connected with the topics discussed in these lecture notes. Schaefer [3], Luxemburg and Zaanen [1] contain excellent references on the theory of ordered vector spaces.

Andô, T.,

1. On fundamental properties of a Banach space with a cone, Pacif. J. Math. 12(1962), 1163-69.

Bauer, T.,

1. Sur le prolongement des formes linéaires positives dans un espace vectoriel ordonne, C.R. Hebd. Séanc. Acad. Sci., Paris 244(1957), 289-92.

Bonsall, F.F.,

1. The decomposition of continuous linear functionals in non-negative components, Proc. Durham Phil. Soc. A13(1957), 6-11.

Brudovskiǐ, B.S.,

1. Associated nuclear topology, mappings of type s , and strongly nuclear spaces, Soviet Math. Dokl. 9(1968), 61-3.
2. s-type mappings of locally convex spaces, ibid. 9(1968), 572-4.

Chaney, J.,

1. Banach lattices of compact maps, Math. Z. 129(1972), 1-19.

Ellis, A.J.,

1. The duality of partially ordered normed linear spaces, J.London Math. Soc. 37(1964) 730-44.

Fremlin, D.H.,

1. Tensor products of Archimedean vector lattices, Amer. J. Math. 44(1972), 777-98.
2. Topological Riesz spaces and measure theory (Cambridge Univ. Press, 1974).
3. Tensor products of Banach lattices, Math. Ann. 211(1974), 87-106.

Husain, T.,

1. The open mapping and closed graph theorems in topological vector spaces (Oxford Math. Monographs, Clarendon Press, Oxford, 1965).

Horváth, J.

1. Topological vector spaces and distributions Vol. 1 (Addison-Wesley Publishing Company, 1966).

Jameson, G.J.O.,

1. Ordered linear spaces (Lecture Notes in Math. 104, Springer-Verlag, Berlin, 1970).

Grothendieck, A.

1. Topological vector spaces (Gordon and Breach, New York, 1973).

Kakutani, S.,

1. Concrete representation of abstract L-spaces and the mean ergodic theorem, Ann. of Math. 42(1941), 523-37.
2. Concrete representation of abstract M-spaces, Ann. of Math. 42(1941), 994-1024.

Kōmura, Y., and S. Koshi,

1. Nuclear vector lattices, Math. Ann. 163(1966), 105-10.

Köthe, G.,

1. Topological vector spaces I (Springer-Verlag, Berlin, 1969).

Luxemburg, W.A.Z. and A.C. Zaanen,

1. Riesz spaces I. (Amsterdam, North Holland, 1971).

May, W.D., and R.R. Chivukula,

1. Lattice properties in $\mathcal{L}(E, F)$, Duke Math. J. (2) 39(1972), 345-50.

Nakano, H.,

1. Linear topologies on semi-ordered linear spaces, J. Fac. Sci. Hokkaido Univ. 12(1953), 87-104.

Namioka, I.,

1. Partially ordered linear topological spaces, Memoirs Amer. Math. Soc. 24(1957).

Ng, Kung-Fu and M.Duhoux, (see also Wong Yau-Chuen)

1. The duality of ordered locally convex spaces, J. London Math. Soc. (2), 8(1973), 201-208.

Peressini, A.L.,

1. Ordered topological vector spaces (Harper-Row, New York, 1967).
2. A note on the lattice properties of continuous linear mappings, Math. Z. 115(1970), 18-22.
3. On topologies in ordered vector spaces, Math. Ann. 144 (1961), 199-223.

Peressini, A.L., and D.R. Sherbert,

1. Ordered topological tensor products, Proc. London Math. Soc. (3) 19(1969), 177-90.

Pietsch, A.,

1. Nuclear locally convex spaces (Springer-Verlag, Berlin, 1972).

Randtke, D.,

1. Characterization of precompact maps, Schwartz spaces and nuclear spaces, Trans. Amer. Math. Soc. 165(1972), 87-101.

Schaefer, H.H.,

1. Topological vector spaces (Springer-Verlag, Berlin, 1971, 3rd print).
2. Normed tensor products of Banach lattices, Israel J. Math. 12(1972), 400-15.
3. Banach lattices and positive operators (Springer-Verlag, Berlin, 1974).

Schlotterbeck, U.,

1. Order-theoretic characterization of Hilbert-Schmidt operators, Arch. Math. 24(1973), 67-70.

Terzioglu, T.,

1. On Schwartz spaces, Math. Ann. 182(1968),236-42.
2. Approximation property of co-nuclear spaces, Math. Ann. 191(1971), 35-37.

Walsh, B.,

1. Ordered vector sequence spaces and related classes of linear operators, Math. Ann. 206(1973), 89-138.

2. An approximation property characterizations ordered vector spaces with lattice-ordered dual, _Bull_. _Amer_. _Math_. _Soc_., 80(1974), 1165-68.

Wittstock, G.,

1. Ordered normed tensor products (In: Foundations of quantum mechanics and ordered linear spaces, 67-84, Lecture Notes in Physis 29, Springer-Verlag, Berlin, 1974).

Wong, Yau-Chuen,

1. On a theorem of Kōmura-Koshi and of Andô-Ellis, _Proc_. _Amer_. _Math_. _Soc_., (to appear).

Wong, Yau-Chuen and W.L. Cheung,

1. Locally absolutely - dominated spaces, _United College J_. (Hong Kong), Vol. 9(1971), 241-49.

Wong, Yau-Chuen and Kung-Fu Ng,

1. _Partially ordered topological vector spaces_ (Oxford Math. Monographs, Clarendon Press, Oxford, 1973).

INDEX AND SYMBOLS

Vol. 399: Functional Analysis and its Applications. Proceedings 1973. Edited by H. G. Garnir, K. R. Unni and J. H. Williamson. II, 584 pages. 1974.

Vol. 400: A Crash Course on Kleinian Groups. Proceedings 1974. Edited by L. Bers and I. Kra. VII, 130 pages. 1974.

Vol. 401: M. F. Atiyah, Elliptic Operators and Compact Groups. V, 93 pages. 1974.

Vol. 402: M. Waldschmidt, Nombres Transcendants. VIII, 277 pages. 1974.

Vol. 403: Combinatorial Mathematics. Proceedings 1972. Edited by D. A. Holton. VIII, 148 pages. 1974.

Vol. 404: Théorie du Potentiel et Analyse Harmonique. Edité par J. Faraut. V, 245 pages. 1974.

Vol. 405: K. J. Devlin and H. Johnsbråten, The Souslin Problem. VIII, 132 pages. 1974.

Vol. 406: Graphs and Combinatorics. Proceedings 1973. Edited by R. A. Bari and F. Harary. VIII, 355 pages. 1974.

Vol. 407: P. Berthelot, Cohomologie Cristalline des Schémas de Caracteristique p > o. II, 604 pages. 1974.

Vol. 408: J. Wermer, Potential Theory. VIII, 146 pages. 1974.

Vol. 409: Fonctions de Plusieurs Variables Complexes, Séminaire François Norguet 1970–1973. XIII, 612 pages. 1974.

Vol. 410: Séminaire Pierre Lelong (Analyse) Année 1972–1973. VI, 181 pages. 1974.

Vol. 411: Hypergraph Seminar. Ohio State University, 1972. Edited by C. Berge and D. Ray-Chaudhuri. IX, 287 pages. 1974.

Vol. 412: Classification of Algebraic Varieties and Compact Complex Manifolds. Proceedings 1974. Edited by H. Popp. V, 333 pages. 1974.

Vol. 413: M. Bruneau, Variation Totale d'une Fonction. XIV, 332 pages. 1974.

Vol. 414: T. Kambayashi, M. Miyanishi and M. Takeuchi, Unipotent Algebraic Groups. VI, 165 pages. 1974.

Vol. 415: Ordinary and Partial Differential Equations. Proceedings 1974. XVII, 447 pages. 1974.

Vol. 416: M. E. Taylor, Pseudo Differential Operators. IV, 155 pages. 1974.

Vol. 417: H. H. Keller, Differential Calculus in Locally Convex Spaces. XVI, 131 pages. 1974.

Vol. 418: Localization in Group Theory and Homotopy Theory and Related Topics. Battelle Seattle 1974 Seminar. Edited by P. J. Hilton. VI, 172 pages 1974.

Vol. 419: Topics in Analysis. Proceedings 1970. Edited by O. E. Lehto, I. S. Louhivaara, and R. H. Nevanlinna. XIII, 392 pages. 1974.

Vol. 420: Category Seminar. Proceedings 1972/73. Edited by G. M. Kelly. VI, 375 pages. 1974.

Vol. 421: V. Poénaru, Groupes Discrets. VI, 216 pages. 1974.

Vol. 422: J.-M. Lemaire, Algèbres Connexes et Homologie des Espaces de Lacets. XIV, 133 pages. 1974.

Vol. 423: S. S. Abhyankar and A. M. Sathaye, Geometric Theory of Algebraic Space Curves. XIV, 302 pages. 1974.

Vol. 424: L. Weiss and J. Wolfowitz, Maximum Probability Estimators and Related Topics. V, 106 pages. 1974.

Vol. 425: P. R. Chernoff and J. E. Marsden, Properties of Infinite Dimensional Hamiltonian Systems. IV, 160 pages. 1974.

Vol. 426: M. L. Silverstein, Symmetric Markov Processes. X, 287 pages. 1974.

Vol. 427: H. Omori, Infinite Dimensional Lie Transformation Groups. XII, 149 pages. 1974.

Vol. 428: Algebraic and Geometrical Methods in Topology, Proceedings 1973. Edited by L. F. McAuley. XI, 280 pages. 1974.

Vol. 429: L. Cohn, Analytic Theory of the Harish-Chandra C-Func tion. III, 154 pages. 1974.

Vol. 430: Constructive and Computational Methods for Differe tial and Integral Equations. Proceedings 1974. Edited by D. L. Colto and R. P. Gilbert. VII, 476 pages. 1974.

Vol. 431: Séminaire Bourbaki – vol. 1973/74. Exposés 436–452 IV, 347 pages. 1975.

Vol. 432: R. P. Pflug, Holomorphiegebiete, pseudokonvexe Ge biete und das Levi-Problem. VI, 210 Seiten. 1975.

Vol. 433: W. G. Faris, Self-Adjoint Operators. VII, 115 pages 1975.

Vol. 434: P. Brenner, V. Thomée, and L. B. Wahlbin, Beso Spaces and Applications to Difference Methods for Initial Value Problems. II, 154 pages. 1975.

Vol. 435: C. F. Dunkl and D. E. Ramirez, Representations of Com mutative Semitopological Semigroups. VI, 181 pages. 1975

Vol. 436: L. Auslander and R. Tolimieri, Abelian Harmonic Analysis Theta Functions and Function Algebras on a Nilmanifold. V, 99 pages. 1975.

Vol. 437: D. W. Masser, Elliptic Functions and Transcendence. XIV, 143 pages. 1975.

Vol. 438: Geometric Topology. Proceedings 1974. Edited by L. C. Glaser and T. B. Rushing. X, 459 pages. 1975.

Vol. 439: K. Ueno, Classification Theory of Algebraic Varieties and Compact Complex Spaces. XIX, 278 pages. 1975

Vol. 440: R. K. Getoor, Markov Processes: Ray Processes and Right Processes. V, 118 pages. 1975.

Vol. 441: N. Jacobson, PI-Algebras. An Introduction. V, 115 pages. 1975.

Vol. 442: C. H. Wilcox, Scattering Theory for the d'Alembert Equation in Exterior Domains. III, 184 pages. 1975.

Vol. 443: M. Lazard, Commutative Formal Groups. II, 236 pages. 1975.

Vol. 444: F. van Oystaeyen, Prime Spectra in Non-Commutative Algebra. V, 128 pages. 1975.

Vol. 445: Model Theory and Topoi. Edited by F. W. Lawvere, C. Maurer, and G. C. Wraith. III, 354 pages. 1975.

Vol. 446: Partial Differential Equations and Related Topics. Proceedings 1974. Edited by J. A. Goldstein. IV, 389 pages. 1975.

Vol. 447: S. Toledo, Tableau Systems for First Order Number Theory and Certain Higher Order Theories. III, 339 pages. 1975.

Vol. 448: Spectral Theory and Differential Equations. Proceedings 1974. Edited by W. N. Everitt. XII, 321 pages. 1975.

Vol. 449: Hyperfunctions and Theoretical Physics. Proceedings 1973. Edited by F. Pham. IV, 218 pages. 1975.

Vol. 450: Algebra and Logic. Proceedings 1974. Edited by J. N. Crossley. VIII, 307 pages. 1975.

Vol. 451: Probabilistic Methods in Differential Equations. Proceedings 1974. Edited by M. A. Pinsky. VII, 162 pages. 1975.

Vol. 452: Combinatorial Mathematics III. Proceedings 1974. Edited by Anne Penfold Street and W. D. Wallis. IX, 233 pages. 1975.

Vol. 453: Logic Colloquium. Symposium on Logic Held at Boston, 1972–73. Edited by R. Parikh. IV, 251 pages. 1975.

Vol. 454: J. Hirschfeld and W. H. Wheeler, Forcing, Arithmetic, Division Rings. VII, 266 pages. 1975.

Vol. 455: H. Kraft, Kommutative algebraische Gruppen und Ringe. III, 163 Seiten. 1975.

Vol. 456: R. M. Fossum, P. A. Griffith, and I. Reiten, Trivial Extensions of Abelian Categories. Homological Algebra of Trivial Extensions of Abelian Categories with Applications to Ring Theory. XI, 122 pages. 1975.

Vol. 457: Fractional Calculus and Its Applications. Proceedings 1974. Edited by B. Ross. VI, 381 pages. 1975.

Vol. 458: P. Walters, Ergodic Theory – Introductory Lectures. VI, 198 pages. 1975.

Vol. 459: Fourier Integral Operators and Partial Differential Equations. Proceedings 1974. Edited by J. Chazarain. VI, 372 pages. 1975.

Vol. 460: O. Loos, Jordan Pairs. XVI, 218 pages. 1975.

Vol. 461: Computational Mechanics. Proceedings 1974. Edited by J. T. Oden. VII, 328 pages. 1975.

Vol. 462: P. Gérardin, Construction de Séries Discrètes p-adiques. »Sur les séries discrètes non ramifiées des groupes réductifs déployés p-adiques«. III, 180 pages. 1975.

Vol. 463: H.-H. Kuo, Gaussian Measures in Banach Spaces. VI, 224 pages. 1975.

Vol. 464: C. Rockland, Hypoellipticity and Eigenvalue Asymptotics. III, 171 pages. 1975.

Vol. 465: Séminaire de Probabilités IX. Proceedings 1973/74. Edité par P. A. Meyer. IV, 589 pages. 1975.

Vol. 466: Non-Commutative Harmonic Analysis. Proceedings 1974. Edited by J. Carmona, J. Dixmier and M. Vergne. VI, 231 pages. 1975.

Vol. 467: M. R. Essén, The Cos $\pi\lambda$ Theorem. With a paper by Christer Borell. VII, 112 pages. 1975.

Vol. 468: Dynamical Systems – Warwick 1974. Proceedings 1973/74. Edited by A. Manning. X, 405 pages. 1975.

Vol. 469: E. Binz, Continuous Convergence on C(X). IX, 140 pages. 1975.

Vol. 470: R. Bowen, Equilibrium States and the Ergodic Theory of Anosov Diffeomorphisms. III, 108 pages. 1975.

Vol. 471: R. S. Hamilton, Harmonic Maps of Manifolds with Boundary. III, 168 pages. 1975.

Vol. 472: Probability-Winter School. Proceedings 1975. Edited by Z. Ciesielski, K. Urbanik, and W. A. Woyczyński. VI, 283 pages. 1975.

Vol. 473: D. Burghelea, R. Lashof, and M. Rothenberg, Groups of Automorphisms of Manifolds. (with an appendix by E. Pedersen) VII, 156 pages. 1975.

Vol. 474: Séminaire Pierre Lelong (Analyse) Année 1973/74. Edité par P. Lelong. VI, 182 pages. 1975.

Vol. 475: Répartition Modulo 1. Actes du Colloque de Marseille-Luminy, 4 au 7 Juin 1974. Edité par G. Rauzy. V, 258 pages. 1975. 1975.

Vol. 476: Modular Functions of One Variable IV. Proceedings 1972. Edited by B. J. Birch and W. Kuyk. V, 151 pages. 1975.

Vol. 477: Optimization and Optimal Control. Proceedings 1974. Edited by R. Bulirsch, W. Oettli, and J. Stoer. VII, 294 pages. 1975.

Vol. 478: G. Schober, Univalent Functions – Selected Topics. V, 200 pages. 1975.

Vol. 479: S. D. Fisher and J. W. Jerome, Minimum Norm Extremals in Function Spaces. With Applications to Classical and Modern Analysis. VIII, 209 pages. 1975.

Vol. 480: X. M. Fernique, J. P. Conze et J. Gani, Ecole d'Eté de Probabilités de Saint-Flour IV–1974. Edité par P.-L. Hennequin. XI, 293 pages. 1975.

Vol. 481: M. de Guzmán, Differentiation of Integrals in R^n. XII, 226 pages. 1975.

Vol. 482: Fonctions de Plusieurs Variables Complexes II. Séminaire François Norguet 1974–1975. IX, 367 pages. 1975.

Vol. 483: R. D. M. Accola, Riemann Surfaces, Theta Functions, and Abelian Automorphisms Groups. III, 105 pages. 1975.

Vol. 484: Differential Topology and Geometry. Proceedings 1974. Edited by G. P. Joubert, R. P. Moussu, and R. H. Roussarie. IX, 287 pages. 1975.

Vol. 485: J. Diestel, Geometry of Banach Spaces – Selected Topics. XI, 282 pages. 1975.

Vol. 486: S. Stratila and D. Voiculescu, Representations of AF-Algebras and of the Group U (∞). IX, 169 pages. 1975.

Vol. 487: H. M. Reimann und T. Rychener, Funktionen beschränkter mittlerer Oszillation. VI, 141 Seiten. 1975.

Vol. 488: Representations of Algebras, Ottawa 1974. Proceedings 1974. Edited by V. Dlab and P. Gabriel. XII, 378 pages. 1975.

Vol. 489: J. Bair and R. Fourneau, Etude Géométrique des Espaces Vectoriels. Une Introduction. VII, 185 pages. 1975.

Vol. 490: The Geometry of Metric and Linear Spaces. Proceedings 1974. Edited by L. M. Kelly. X, 244 pages. 1975.

Vol. 491: K. A. Broughan, Invariants for Real-Generated Uniform Topological and Algebraic Categories. X, 197 pages. 1975.

Vol. 492: Infinitary Logic: In Memoriam Carol Karp. Edited by D. W. Kueker. VI, 206 pages. 1975.

Vol. 493: F. W. Kamber and P. Tondeur, Foliated Bundles and Characteristic Classes. XIII, 208 pages. 1975.

Vol. 494: A Cornea and G. Licea. Order and Potential Resolvent Families of Kernels. IV, 154 pages. 1975.

Vol. 495: A. Kerber, Representations of Permutation Groups II. V, 175 pages. 1975.

Vol. 496: L. H. Hodgkin and V. P. Snaith, Topics in K-Theory. Two Independent Contributions. III, 294 pages. 1975.

Vol. 497: Analyse Harmonique sur les Groupes de Lie. Proceedings 1973–75. Edité par P. Eymard et al. VI, 710 pages. 1975.

Vol. 498: Model Theory and Algebra. A Memorial Tribute to Abraham Robinson. Edited by D. H. Saracino and V. B. Weispfenning. X, 463 pages. 1975.

Vol. 499: Logic Conference, Kiel 1974. Proceedings. Edited by G. H. Müller, A. Oberschelp, and K. Potthoff. V, 651 pages 1975.

Vol. 500: Proof Theory Symposion, Kiel 1974. Proceedings. Edited by J. Diller and G. H. Müller. VIII, 383 pages. 1975.

Vol. 501: Spline Functions, Karlsruhe 1975. Proceedings. Edited by K. Böhmer, G. Meinardus, and W. Schempp. VI, 421 pages. 1976.

Vol. 502: János Galambos, Representations of Real Numbers by Infinite Series. VI, 146 pages. 1976.

Vol. 503: Applications of Methods of Functional Analysis to Problems in Mechanics. Proceedings 1975. Edited by P. Germain and B. Nayroles. XIX, 531 pages. 1976.

Vol. 504: S. Lang and H. F. Trotter, Frobenius Distributions in GL_2-Extensions. III, 274 pages. 1976.

Vol. 505: Advances in Complex Function Theory. Proceedings 1973/74. Edited by W. E. Kirwan and L. Zalcman. VIII, 203 pages. 1976.

Vol. 506: Numerical Analysis, Dundee 1975. Proceedings. Edited by G. A. Watson. X, 201 pages. 1976.

Vol. 507: M. C. Reed, Abstract Non-Linear Wave Equations. VI, 128 pages. 1976.

Vol. 508: E. Seneta, Regularly Varying Functions. V, 112 pages. 1976.

Vol. 509: D. E. Blair, Contact Manifolds in Riemannian Geometry. VI, 146 pages. 1976.

Vol. 510: V. Poènaru, Singularités C^∞ en Présence de Symétrie. V, 174 pages. 1976.

Vol. 511: Séminaire de Probabilités X. Proceedings 1974/75. Edité par P. A. Meyer. VI, 593 pages. 1976.

Vol. 512: Spaces of Analytic Functions, Kristiansand, Norway 1975. Proceedings. Edited by O. B. Bekken, B. K. Øksendal, and A. Stray. VIII, 204 pages. 1976.

Vol. 513: R. B. Warfield, Jr. Nilpotent Groups. VIII, 115 pages. 1976.

Vol. 514: Séminaire Bourbaki vol. 1974/75. Exposés 453 – 470. IV, 276 pages. 1976.

Vol. 515: Bäcklund Transformations. Nashville, Tennessee 1974. Proceedings. Edited by R. M. Miura. VIII, 295 pages. 1976.